Theories
of Local
Economic
Development

The two editors of this book have historically approached economic development from very different perspectives. Robert Mier's approach has always been community based. His focus has been on developing jobs and wealth in the city's neighborhoods. His goal has always been to improve the conditions of neighborhood residents through economic development—particularly the conditions of those in poor neighborhoods.

Richard Bingham is an unabashed empiricist. He is interested in the location and movement of jobs and industries between and within regions. He is interested in describing and explaining empirically verified changes in the locational patterns of industries and in explaining and predicting the development and growth of agglomerations such as edge cities.

How could two individuals with such apparently unrelated interests join together to edit *Theories of Local Economic Development: Perspectives From Across the Disciplines?* Through casual conversation both found that they had one important life experience in common—service in Vietnam. The Vietnam War is truly a common bond that transcends ideologies. From this common bond, a professional and personal friendship developed in which both found their philosophies similar but their approaches different—hence this book; hence this dedication:

TO THE VETERANS OF VIETNAM

Theories
of Local
Economic
Development

Perspectives From Across the Disciplines

edited by

Richard D. Bingham
Robert Mier

SAGE Publications
International Educational and Professional Publisher
Newbury Park   London   New Delhi

For information address:

SAGE Publications, Inc.
2455 Teller Road
Newbury Park, California 91320

SAGE Publications Ltd.
6 Bonhill Street
London EC2A 4PU
United Kingdom

SAGE Publications India Pvt. Ltd.
M-32 Market
Greater Kailash I
New Delhi 110 048 India

Printed in the United States of America

Library of Congress Cataloging-in-Publication Data

Main entry under title:

Theories of local economic development : perspectives from across the
  disciplines / edited by Richard D. Bingham, Robert Mier.
      p. cm.
    Includes bibliographical references and indexes.
    ISBN 0-8039-4867-0. — ISBN 0-8039-4868-9 (pbk.)
    1. Industrial promotion—United States. 2. United States—
  Economic policy—1981- 3. Local government—United States.
  I. Bingham, Richard D. II. Mier, Robert.
  HC110.I53T48   1993
  338.973—dc20   93-17611

93 94 95 96 10 9 8 7 6 5 4 3 2 1

Sage Production Editor: Astrid Virding

Contents

Preface

Economic development has been defined by the American Economic Development Council (AEDC) as "the process of creating wealth through the mobilization of human, financial, capital, physical and natural resources to generate marketable goods and services" (AEDC, 1984: 18). Yet the definition from economic development's oldest professional association ignores many important issues such as the roles of the public and private sectors in the process of creating wealth and, more important, how that wealth is distributed. We agree with the AEDC that economic development is about the process of creating jobs and wealth. But it is the role of the private sector to create wealth by producing tradable goods and services and engaging in these exchanges. It is the role of the public sector to facilitate and promote the creation of jobs and wealth by the private sector, *and to ensure that it does so in a way that serves the short- and long-run interests of the broad population* (adapted from Bendavid-Val, 1991: 21).

This is not to suggest that the private sector is the only creator of wealth. Government, and to some extent the nonprofit sector, create wealth in numerous ways—through research and development, investment in infrastructure, and even in the provision of services (e.g., launching telecommunication satellites). It is, however, generally not the role of government in the United States to directly produce tradable goods (munitions, perhaps, being an exception).

Economic development as a public sector activity in the United States has been around for a long time. It effectively began in 1937 in Mississippi with the issuance of the first industrial development bond. Donald Haider (1989) identifies four eras of changing approaches to economic development practice characteristic of the past. The first ran from the early development of location incentives through the early 1960s. In this phase of industrial recruitment, states attempted to attract manufacturing plants from other states to redistribute employment. Federal economic development efforts of the period also sought to stimulate economic development

AUTHORS' NOTE: We wish to thank Robert Beauregard for constructive comments on the Preface.

vii

in depressed areas. For example, in 1961 the Area Redevelopment Administration (precursor to the Economic Development Administration) was established in the Department of Commerce to provide technical assistance, loans, and grants to local governments for public projects—particularly those that would attract private businesses.

During the 1960s the economic development goal was to improve equity and increase demand through redistribution. The federal, state, and local programs of this era sought to provide subsidies to individuals and regions in poverty as a way both to raise income immediately and to stimulate increased economic activity by increasing consumer demand. A major emphasis was on broadening economic opportunity through education, job training, social services, and community development.

Government's contribution to economic revitalization during the 1970s consisted of programs that attempted to stimulate economic development in depressed areas and declining city neighborhoods through a combination of industrial recruitment and private sector investments. The key term was *public-private partnerships*—the conditioning of government assistance to significant private participation in government-approved projects. Public sector interest was in the generation of employment and jobs. Given the relative decline in manufacturing vis-à-vis services during the decade, much of the program emphasis was on service-related employment through the development of hotels, office buildings, and other real estate-related activities, especially in urban areas that had experienced a significant loss of jobs and population. The federal Urban Development Action Grant (UDAG) program is characteristic of this decade of development (see Rich, 1992, for a thorough description).

Finally, the 1980s is a decade of generative development. The controversial work of David Birch (1978) suggested that small businesses, as opposed to large, were the real generator of new jobs. Thus governments shifted their focus to entrepreneurship and assistance to small and medium-sized firms. Governments subscribed to the philosophy that policy should not be aimed toward creating jobs but toward facilitating the enhancement of market mechanisms to create wealth, which, in the process, would create jobs (Haider, 1989).

Robert Mier and Joan Fitzgerald (1991) identify three phases of the development of the scholarly literature in economic development—periods not necessarily coinciding with professional practice. The first began with state efforts at attracting industry in the South in the 1930s. The development practice subsidized "smokestack chasing" with tax abatements, loan packaging, infrastructure and land development, and other efforts that would reduce the cost of production of firms. Two distinct literatures emerged to explain these development practices—"a regional and community development literature borrowing heavily from international development theories and experiences, and an industrial location literature borrowing heavily from theories of firm behavior" (p. 269).

The second phase in the literature on local economic development began to emerge in the late 1960s. This literature arose from concerns about the distribution of benefits from economic development practice. This analysis of the political economy of development was largely a critical one and was based on a Marxist framework.

The third phase of the literature is reflective of the birth of the modern public-private partnerships. The emphasis is now on the promotion of development from within, to reduce local dependence on nonlocal corporations and to broaden the benefits of development to more groups within the locality. Overall, then, Mier and Fitzgerald (1991) see this literature as "the latest of a three-decade academic effort to advance the practice of local economic development *while flirting with the establishment of a new academic discipline*" (pp. 268-269; emphasis added).

Our purpose here is not to declare economic development an academic discipline but to continue the flirtation. One of the hallmarks of an academic discipline is the presence of theory or theories, although not everyone believes that good analysis involves general laws. The pivotal assumption behind this book, however, is that research should and can lead to general statements about economic development. This assumption implies that the behavior of firms and the development process can be explained and predicted in terms of general laws established by observation.

The accumulation of knowledge consists of the process of gradual confirmation and/or modification of the theories that serve as the general premises in the explanatory scheme. We expect much of theories. First, we expect a theory to be accurate—that is, to explain a phenomenon as completely as possible and to predict as much variation as possible. We also expect a theory to be both parsimonious and generalizable. But there are trade-offs. When the accuracy of a theory is maximized, its generality and parsimony will often be low. Finally, the fourth criterion imposed on theory is causality. In general, the extent to which a theory is causal increases as the number of factors incorporated into the theory increases (Przeworski and Teune, 1970: 17-23).

At any stage of development of science, more than one theory will explain the same events. For us this is fortunate. Given the fact that economic development draws on a wide variety of disciplines, the field is not suffering from a lack of theory. To the contrary, so many theories apply that the field of economic development is a very confusing place indeed.

We hope this book will alleviate some of the confusion. It is our goal to bring together theories from a wide variety of disciplines that apply to local economic development. The chapters that follow draw on theories of development from economics, business administration, regional science, planning, political science, public administration, and psychology, to name a few.

The challenge placed before the authors was a formidable one. Each was asked to explore various theories of explanation and prediction emerging from social science disciplines and to focus them on the practical actions of local economic development. The language of scholarship is precise, clearly bounded, and objective. That of practice, however, is fluid, subjective, unstable, contradictory, and often paradoxical. Yet, in every chapter the authors present illustrative case material in an attempt to bridge this language gap and to find evidence of their theory present in the practical work of local economic development.

The book is divided into six parts based upon the different theoretical perspectives presented. Part I, "Location and Space Theories," is composed of two chapters that discuss theoretical perspectives on local economic development representative of the disciplines of business administration, economics, geography, regional science, and to some extent urban planning.

The second part is "Space-Based Strategies." Each chapter is concerned with specific spatial dimensions of local economic development—rural development, neighborhood development, and ghetto development. The three chapters in this part draw on the disciplines of geography, urban planning, sociology, and political science.

There are two chapters in the third part, "Labor and Capital Theories." The perspectives presented draw on the disciplines of economics, sociology, and urban planning.

Two chapters represent "Political and Social Theories." The chapters largely present perspectives on local economic development as seen by political economists and urban planners.

The final substantive part of the book examines "Organization and Process." The three chapters concern citizenship and local economic development, technology transfer, and entrepreneurship. The theories presented are largely drawn from the disciplines of political science, public administration, and economics.

The concluding two chapters constitute the book's "Theoretical Perspectives." The first exposes the constitutive rules of local economic development as practiced in the United States, and the second attempts to connect theory to practice through a framework of alternative metaphors.

The first chapter, by John Blair and Robert Premus, "Location Theory," discusses economic development in terms of the theory of economics and business administration. As they point out: "Location theory in economics evolved from simple transportation cost minimization models." Later these models were replaced by more complex locational models incorporating numerous other locational determinants within a cost of production framework. These include labor costs, productivity, the labor environment, industrial filtering, taxes and business climate, government, politics, amenities, and quality of life. Blair and Premus discuss the concept of inertia—the tendency to stay put—as an important locational concept. They then discuss agglomeration economies and coevolutionary development (the development of symbiotic relationships in a locality). The chapter concludes with a critical evaluation of location theory.

Chapter 2, "Theories of Regional Development," by Arthur Nelson, draws its theory from the fields of geography and regional science. Nelson classified theories of regional development into two schools, both having as a common ground the application of economic base theory.[1] They are the *development-from-above school*, which views development as emanating from the core and spreading to the periphery and hinterland, and the *development-from-below school*, which argues for regions to take control of their own institutions.

The development-from-above school is characterized by price equilibrium models of regional development, the dynamic disequilibrium model, product life-cycle theory, and regional life-cycle models. Development-from-below aims for generative growth and is exemplified by territorial development theory, functional development theory, and agropolitan development theory. Nelson concludes with an assessment of the determinants of manufacturing location between core and peripheral regions, and between centers and their hinterlands, and presents implications for regional development professionals.

Chapter 3, "Applying Theory to Practice in Rural Economies," is by Marie Howland. She notes that rural economies tend to be small, specialized, and disproportionately composed of a low-skilled labor force. Thus agglomeration economies are generally absent. The theories applying to rural economic development include central place theory, product cycle theory, the circular and cumulative causation model, and both demand- and supply-side theories. The chapter concludes with three case examples of economic theory applied to rural development. The cases cover rural downtown development, rural manufacturing, and a flourishing tourism/retirement community.

In Chapter 4, "The Economic Development of Neighborhoods and Localities," Wim Wiewel, Michael Teitz, and Robert Giloth discuss the theory and practice of neighborhood-based economic development. As the authors point out: There are four bodies of theoretical work relevant to neighborhood economic development. These are market-based theories of economic development, Marxian critiques of these theories, theories of the political process, and sociological theories of community. Wiewel, Teitz, and Giloth then apply the theories to eight forms of neighborhood economic development practice: business retention, commercial revitalization, new business ventures, entrepreneurialism, neighborhood capital accumulation, education and training, labor-based development, and community organizing/planning. This then forms the basis for the development of criteria for a new theory.

Chapter 5, "Ghetto Economic Development," by William Goldsmith and Lewis Randolph, focuses on a movement that gained impetus along with the civil rights struggles of the 1960s and has experienced a resurgence of interest since the Los Angeles disturbances of 1992. They trace the historical struggle of black capitalism waged for a century within the African American community beginning with debates between W. E. B. DuBois and Booker T. Washington. Following assessment of the theory and practice of black capitalism, they delve into international development theory to examine critically how those theories have been applied to efforts to attack a supposed "culture of poverty." Finally, they bring up to date the policy debate on dispersing the ghetto. All this is used as a framework for reviewing four current policy directions for promoting ghetto economic development.

In Chapter 6, "Labor Force, Education, and Work," Joan Fitzgerald reviews the impact of these theories on local economic development. She describes and analyzes the skills mismatch debate, training as economic development, and education as economic development. She then presents two cases as examples of how education and training reform informed by an equity model of work force education and training might look. Fitzgerald concludes that the traditional corporatist approaches do not address the inherently unequal distribution of education and training funds; however, the cases illustrate that blue-collar workers and inner-city residents can participate in defining these problems and shaping effective solutions.

Chapter 7, "Theory and Practice in High-Tech Economic Development," is by Harvey Goldstein and Michael Luger. By *high tech* they mean those industrial organizations "engaged in technically sophisticated activities that lead to product or process innovations, new inventions, or, more generally, the creation of knowledge." Theories related to high-tech economic development include the neoclassical economic theories: regional growth theory, comparative advantage, disequilibrium models, growth pole theory, Weberian location theory, and innovation theory. They also include cyclical theories: stages theory, long wave theories, and product cycle theory. The final group of theories discussed by Goldstein and Luger are theories of propulsive, innovative, and creative regions. They then develop a typology of policy approaches based in practice, identify specific programs that fit within it, and discuss the two-way relationship between theory and policy.

Chapter 8, "Political Economy and Urban Development," by Scott Holupka and Anne Shlay, draws its theory from a political economy framework. The political economy approach, sometimes described as the institutional approach or the "new urban theory," emphasizes how the actions and behaviors of key institutions and actors are responsible for determining how urban structures develop. Urban development outcomes viewed as "inevitable" under the traditional perspective are seen as being contingent upon specific actions taken by specific actors and institutions in this new theoretical framework. Theoretical developments in economic development stemming from the political economy tradition include political economic theory and the thesis that a city is a "growth machine" (Molotch, 1976, 1979). The chapter concludes with several excellent case studies.

In Chapter 9, "Race and Class in Local Economic Development," John Betancur and Douglas Gills confront this thorny issue. They first consider four political and policy tensions that create the context within which low-income and minority people pursue economic development. These are the tension between nationality-based strategies versus ones that are "color blind," one between community-based strategies versus geographic dispersal, one between nationality-controlled development versus majority controlled, and finally one between overtly equity-oriented development versus growth and "trickle down"-oriented development. They next review three dominant theories of development molded by these tensions: community development, affirmative action, and separatism. They conclude by lamenting that the class implications of development, in the contemporary United States, are overwhelmed by the race dynamic, making it necessary for any strategy confronting the inequality of wealth and opportunity to face the racial divisions in our society.

The tenth chapter, "Citizenship and Economic Development," by Elaine Sharp and Michael Bath, draws on the citizen participation theories of political science. Sharp and Bath point out that applying theories of political participation to citizen involvement in local economic development is no easy task. "First, most theories point to what it is that activates citizens rather than what predicts the direction (valence) of their activation or the quality and effects of their participation," and, second, some suggest "that citizen participation in the developmental sphere will tend to be limited, especially by comparison to citizen involvement in the distributional arena." The authors point out that case study evidence suggests that the distinctive characteristic of citizen participation with respect to economic development matters is its group-based orientation. A number of influential theories have been developed that account for the various forms of political participation that are relevant to development decisions. They fall into three categories. The first are the psychosocial theories, which emphasize individual attitudes, beliefs, and so on, and the social groupings that condition their development. The second are rational calculus theories, which assume that individuals are mobilized into group-based political action on the basis of their objective assessments of the impact of the proposed policies. The third, institutionalist theories, emphasizes the importance of institutional arrangements in either fostering or limiting citizens' access to governmental decision making. Then, through the use of illustrative case studies, Sharp and Bath show that the dynamics of citizen participation in economic development are substantially different in economically distressed and economically advantaged communities.

Chapter 11, "Technology Transfer and Economic Development," is by Julia Melkers, Daniel Bugler, and Barry Bozeman. The authors focus on innovation theories and technology transfer. They conceptualize technology transfer in terms of three models: appropriability, dissemination, and knowledge use. The chapter concludes with a discussion relating transfer theories to different types of state and local development programs.

The final substantive chapter is Chapter 12, "Theories of Entrepreneurship," by Timothy Bates, who emphasizes economic and sociological paradigms as they apply to entrepreneurship and development. Economists traditionally emphasize the monetary rewards of entrepreneurship relative to alternative employment in explaining the decision to pursue self-employment. Sociologists, on the other hand, explain entrepreneurship by focusing upon the social environment that encompasses cultural as well as economic factors. Bates discusses theories of entrepreneurship as they pertain to monetary rewards, entrepreneurial ability, financial support, cultural factors, protected markets, and ethnic enclaves. Bates concludes that "a consistent set of traits broadly describes those who are most likely to enter self-employment and to operate businesses that are likely to survive."

The final two chapters of the book attempt to view what has been discussed in the previous chapters in a larger perspective. Chapter 13 by Robert Beauregard, "Constituting Economic Development: A Theoretical Perspective," exposes the constitutive rules of economic development as practiced in the United States. "Constitutive rules are the seldom considered assumptions and theoretical relationships that direct our thinking, and failure to critically evaluate them hides ideological biases and suppresses the inevitable tensions and disagreements that arise in a multiethnic society of tenacious inequalities, precarious democratic practices, and deeply ingrained capitalist values." Beauregard probes the categories of economic development and the epistemology used to fill them, that is, the rules we use to distinguish valid from invalid evidence and procedures. He characterizes economic development in the United States as investor centered, tied to the core capitalist institutions, antithetical toward history, temporally unsophisticated, and spatially conflicted. Epistemologically, economic development officials struggle with the incompatibility of law-like quantitative analyses and particularistic stories that tout successes and opportunities. "What knowledge is declared valid, relevant, or useful depends on which actors voice that knowledge," however. Beauregard concludes that economic development "involves a complex web of coexisting and conflicting knowledge bases. Practice vacillates between objectivist and subjectivist epistemologies, with the contradictions often resolved in favor of privileged actors."

Finally, in the last chapter, "Metaphors of Economic Development," we recognize the folly of our initial goal in undertaking this book—that of articulating a synthetic theory of local economic development. This never occurred, as the number and complexity of the theories and models covered in the first 12 chapters made synthesis an impossible task. The authors of these chapters discussed and analyzed more than 50 theories drawn from the various social sciences that pertain directly to economic development.

Instead, we embarked on another course: We view the theories and cases presented by the contributing authors within a framework of metaphors that lead us to understand economic development in distinctive yet partial ways. The premise of the concluding chapter is that the theories of economic development presented in the balance of the book must be viewed within frameworks that force the development scholar or practitioner to engage with the multiple meanings of development rather than pretending they don't exist. Drawing on the work of Mier and Fitzgerald (1991), we argue that generative metaphors are essential to this task of confronting ambiguity and thus a key to the integration of scholarship and professional practice. We suggest seven metaphorical ways of reading and seeing local economic development emerge from a consideration of the theories, models, and cases reviewed in the book. They are as follows:

- Economic development as problem solving
- Economic development as running a business
- Economic development as building a growth machine
- Economic development as preserving nature and place
- Economic development as releasing human potential
- Economic development as exerting leadership
- Economic development as a quest for social justice

We conclude by suggesting a means to use these metaphors to construct stories of local economic development. Much of the practice of local development, we argue, is the promotion of alternative, better futures. We suggest that alternative visions require a new social construction of reality—new patterns of perception embedded in narrative habits and patterns of seeing. So inspirational story telling, we believe, takes priority in economic development policy and strategy formulation. Metaphors provide the inspirational spine to the stories and thus a framework for incorporating the diverse theories of development into practical applications. We hope we have been successful in suggesting a practical framework for using theory in local economic development.

Note

1. An excellent review of research on the economic base model was recently completed by Krikelas (1992).

References

American Economic Development Council (AEDC) (1984). *Economic Development Today.* Chicago: AEDC.

Bendavid-Val, Avrom (1991). *Regional and Local Economic Analysis for Practitioners* (4th ed.). New York: Praeger.

Birch, David L. (1978). *The Job Generation Process.* Cambridge: MIT Press.

Haider, D. (1989). "Economic Development: Changing Practices in a Changing US Economy." *Environment and Planning C* 7 (November): 451-469.

Krikelas, A. C. (1992). "Why Regions Grow: A Review of Research on the Economic Base Model." *Economic Review* 77 (July/August): 16-29.

Mier, Robert and Joan Fitzgerald (1991). "Managing Economic Development." *Economic Development Quarterly* 5 (August): 268-279.

Molotch, H. L. (1976). "The City as a Growth Machine." *American Journal of Sociology* 82: 309-330.

Molotch, H. L. (1979). "Capital and Neighborhoods in the United States: Some Conceptual Links." *Urban Affairs Quarterly* 14: 289-312.

Przeworski, Adam and Henry Teune (1970). *The Logic of Comparative Social Inquiry* (Reprint). Malabar, FL: Krieger.

Rich, M. J. (1992). "UDAG, Economic Development, and the Death and Life of American Cities." *Economic Development Quarterly* 6 (May): 150-172.

PART I

Location and Space Theories

1

Location Theory

JOHN P. BLAIR
ROBERT PREMUS

Location theory in economics evolved from simple transportation cost minimization models. As the theory progressed, spatial variations in market size, production cost differentials, regional amenities, technological capabilities, and other factors were integrated into increasingly complex models of the industrial location decision process. These extensions, which paralleled empirical studies on the importance of locational factors, resulted in significant advances in location theory, but paradoxically they also created less deterministic theoretical models of where establishments actually choose to locate. In the real world, inertia, evolutionary processes, and serendipity appear to exert an important influence on industrial location patterns.

In general, studies prior to 1960 indicated that basic cost factors were the dominant determinants of industrial location patterns. Of primary importance were transportation, access to markets, access to material inputs, and the availability (and cost) of labor. More recent empirical studies indicate that, as the economy and technology became more complex, the list of significant locational factors has been lengthened. In addition to the basic cost factors, technical competence of the labor force, state and local taxes, regional business climates, quality of life factors, and other regional differences have been found to exert an influence on business location decisions. Thus, while traditional basic cost factors continue to be the most important locational criteria in many industries, they have declined in relative importance as the other locational factors became more important (Blair and Premus, 1987).

This review begins with an explanation of one of the simplest location models: the transportation cost minimization case. The analysis is then extended to discuss how spatial variations in productions costs and other locational determinants can be incorporated into the transportation cost

3

framework. The third and fourth sections discuss the implications of inertia, agglomeration economies, and coevolutionary processes in location theory development. A case study is then used to illustrate how some of these factors operated in the location of an advanced technology facility. The final section concludes the chapter with a critical appraisal of the current state of location theory.

Transportation Cost Minimization Models

Historically, transportation has been considered one of the dominant locational factors (Thisse, 1987: 519-521). Firm output and raw material inputs are two types of commodities that normally require transport. Hence low transportation costs are integrally linked with two other locational factors: access to markets and access to materials.

Transportation cost models are particularly useful for explaining locational outcomes of establishments producing items that have high shipping costs for either inputs or outputs relative to the value of the final product. Activities that are sensitive in their location decisions to transportation costs are called "transportation oriented."

A One-Input, One-Market Model

Transportation cost models are part of the "tool kit" economists use in analyzing the industrial location decisions of firms. The simplest models assume one market and one input source. Other simplifying assumptions are that costs of production are uniform over space and the volume of the firm's sales are unaffected by the choice of location. Under these circumstances, profit maximizing firms will choose from among a set of location options the site that will minimize total transportation costs.

Figure 1.1 illustrates the location decision under these simplifying assumptions. The firm may locate at any point along a line between the input site and the market site. The cost of transporting the input to the market is called the "assembly cost." "Distribution costs" are the costs of shipping the finished product to the market. In Figure 1.1, both the assembly and the distribution costs are assumed to be linear, implying that total assembly and total distribution costs are proportional to distance from raw materials and market sites, respectively. Long haul economies are assumed not to exist in this version of the model.

Wherever the establishment decides to locate, its total transportation costs will be the sum of the assembly and distribution costs. For example, a location at point Q will result in total transportation costs of $9 (assembly

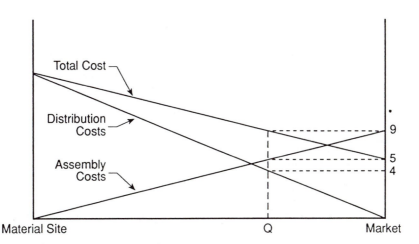

Figure 1.1. A Transportation Cost Minimizing Model

cost, $5; distribution costs, $4). The transportation cost minimizing point will be at the market in this example.

Ideal weights. Locational theorists use the concept of ideal (or locational) weights to help evaluate the influence exerted on the locational decisions of firms by their need to have access to raw material sources and output to markets. The ideal weight of the output is the cost of shipping one unit of output one mile or some other appropriate measure of distance. The ideal weight of the input is the cost per mile of shipping enough of the input to produce one unit of output.

If the output is expensive to ship compared with the input, the activity will be market oriented. The profit maximizing firm will transport its inputs to the market to avoid the higher cost of shipping the output. Conversely, if the ideal weights of the inputs are high compared with the costs of transporting the output, manufacturing will occur at the input site.

Weight-losing production processes tend to be oriented toward the inputs. A milling operation will have a tendency to locate near a timber supply because weight is lost in the milling process. In addition to weight-losing processes, activities tend to locate near input sources when the inputs they need are bulky, heavy, fragile, hazardous, or otherwise expensive to transport.

Market-oriented activities tend to have final outputs (products) that are hazardous to transport, bulky, perishable, or fragile. Also, products that gain weight during the production process by the addition of a resource that is readily available everywhere (i.e., ubiquitous inputs) tend to be market oriented in their location choices. For instance, soft drinks are considered weight-gaining products because water, a ubiquitous input, is added to the syrup and bottle. If production took place away from the market site, the water—the heaviest component—would have to be shipped from the bottling plant to the market.

End-Point Locations

The simple locational model just described implies that firms will find a location at either their input or the market to be more profitable than any of the possible midpoint locations. Total transportation costs will be minimized by shipping either the input or the product, whichever has the lower ideal weight. A site at some midpoint location between markets and the raw material source would be expected only if the ideal weights of the input and output were equal. If this were to happen, the location decision would be indeterminant because the firms would be indifferent to any location along the line.

The tendency of firms to locate at one of the end-point locations is enhanced by two additional features of the transportation system. First, any midpoint location will entail extra loading and unloading costs. This additional factor will increase the locational attractiveness of end-point locations. A second factor that enhances the locational pull toward end points is the presence of long haul economies in transportation. Long haul economies occur when shipment costs per mile decrease the longer the distance shipped. For example, it is frequently more economical to ship one long haul rather than transport a product the same distance in two or more short hauls. Shipping either the product or the input from an end-point site would capture these economies, thus reinforcing the tendency of transportation-oriented firms to locate at one of the end points in the model.

An important exception to the tendency to avoid midpoint locations occurs when there are natural interruptions in the transportation grid. For instance, water ports and railroad terminals have been the site of some manufacturing activities. Because a disruption in the shipment route occurs at these points, many producers have found them to be convenient low-cost location sites for their production facilities, given that loading and unloading would be necessary anyway. Cities such as St. Louis, New Orleans, Buffalo, and Chicago gained prominence as manufacturing sites, in part, because of their location at points of disruption in the transportation grid.

Multiple Inputs

The concept of ideal, or locational, weights also helps explain where facilities will locate when there is more than one input. For instance, suppose two inputs are used in the production process, the firm produces a single output, and the transportation system is dense enough so that it may be approximated by a uniform transportation surface. In this case, the locational weights will have pulls proportional to actual weights. If one of the locational weights is greater than the other two combined (called a "dominant weight"), the optimum location for the firm will be at the site of the dominant input or the market. If neither input has a dominant ideal weight, however, the optimum location site will be where the weights achieve a physical balance. At point t in Figure 1.2(a), any location closer to the market site will result in a higher cost of shipping the two inputs than would be saved from shipping the firm's output a shorter distance to market. Similarly, any location closer to either of the inputs would result in a net increase in total transportation costs, because the sum of the ideal weights being moved toward an input source would exceed the ideal weight of the input with the reduced transportation cost.

As the number of input sources and the number of markets increases, the transportation cost minimizing model becomes more complex, but the concept of locational weights remains useful. Consider Figure 1.2(b), which shows three inputs and one market. A transportation grid is shown so that transportation can take place only along the given transportation system. The optimal location is at "t." Any movement away from "t" will result in increased transportation costs.

The transportation cost model suggests that if the output has a dominant locational weight the establishment will locate at the market. In the previous example, the market was considered a point and there was only one market. The market can be more realistically portrayed as a metropolitan area consisting of a number of neighborhoods. Thus it would be appropriate to ask: Where within a given market area would a transportation-oriented establishment locate?

Retail activity and the principle of median location. The concept of ideal weights can also be used to explain the location of retail activities. For most retail activities, the customer can be thought of as the factor that needs to be transported to the site of purchase. Customers will shop, possibly make a purchase, and then transport the product back to their residence. In this case, retail activities tend to be market oriented because

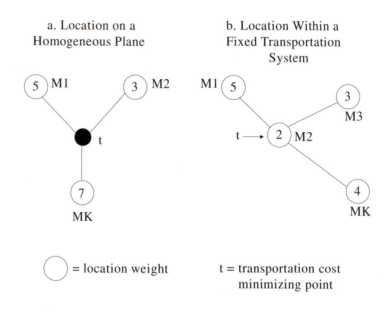

= location weight   t = transportation cost
                              minimizing point

Figure 1.2. Transportation Costs and Multiple Inputs

people are very expensive to transport. Not only do customers have high
opportunity costs, but they must travel in safety and relative comfort.

The principle of median location is useful in explaining the location of
market-oriented activities such as retail or wholesale market activities.
Suppose a producer is considering where to locate along a line, as illustrated in Figure 1.3. The number of customers at each location is indicated
at points on the line. The model assumes that the firm will make deliveries
to customers at random times and that each delivery requires one round
trip. Bundling of trips is not allowed. Under these circumstances, where
will the transportation cost minimization site be located?

Many observers initially select point f as the transportation cost minimizing location because it is in the middle of the market being served.
Closer analysis, however, indicates that "g" is the transportation cost
minimization location. To understand why "g" is the transportation cost
minimization location for a market-oriented product, suppose a firm located at "g" considered moving one "block" to the east (right). The number
of block-trips saved serving eastern customers would be 10. But the
number of extra block trips required to serve western customers would
increase by 12. The principle of median location also applies to distribution centers serving several cities as well as retail establishments serving

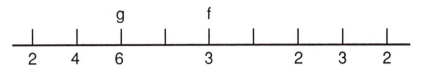

Figure 1.3. The Principle of Median Location

neighborhoods. Accordingly, a firm distributing to 100 retailers in city A, 200 retailers in city B, and 700 retailers in city C will tend to locate in city C. Because the medium location of customers will tend to be in the largest cities, large cities will have an advantage as distribution and service centers. Similarly, the most densely populated neighborhoods within cities, such as the central business district, will benefit from the principle of medium location.

This discussion of the transportation cost minimization model illustrates the key principles upon which industrial location theory has evolved. This body of location theory serves as the basis in the next section for discussing more complex—albeit less deterministic—models of the industrial location decision process.

Production Costs

Transportation costs have traditionally received more attention than other locational factors for several reasons. First, economic and mathematical models are applicable to solving cost minimization problems. Second, when location theories were initially being developed, many analysts implicitly assumed they were modeling the location decisions of manufacturing establishments, which tend to be transportation oriented. Thus earlier researchers tended to focus on transportation cost minimization models.

Today, the simple transportation cost minimization models have been replaced by more complex location models that incorporate numerous other possible locational determinants. Transportation costs have fallen dramatically over past decades. In addition, advances in product and process technologies have resulted in a shift to high value added activities in the manufacturing and service sectors. Other things equal, the higher the value of the product being shipped relative to shipping costs, the less importance businesses will attach to shipping costs in their location decision. Thus, as industry became more "footloose" in location choices (firms are less influenced by basic cost factors), the range of possible locations expanded as did the list of possible locational determinants. In the case of location theory, the cost of increased realism was a reduction in generality (Stevens, 1985).

Figure 1.4 illustrates one way to incorporate production costs into the transportation cost minimizing model. MTCP represents the minimum transportation cost point. The progressively larger circles around MTCP represent increases in transportation costs. For instance, the circle labeled 1 indicates the points at which transportation costs are $1 greater than at MTCP. Notice that transportation costs rise with distance from MTCP. At point L, the transportation costs are $3 per unit higher than they would be at MTCP.

Suppose a low-cost production point is located at "L." The low production costs may be due to lower costs of a nontransferable input, such as labor, local public services, or taxes. The firm would find it profitable to locate at "L" if their per unit production costs are reduced by more than $3 per unit.

Labor Factors

Access to and cost of labor have been a traditionally important locational factor. The importance of labor costs is obvious because of the high percentage of production costs directly attributable to labor. For instance, wages constitute about 60% of gross national product. It is, however, not always easy to determine which geographic areas offer the lowest labor costs per unit of output.

Prevailing wages. The prevailing, or average, wage is the most common indicator of labor costs. There are, however, several considerations that may drive a wedge between the prevailing wage and a firm's unit labor costs. First, if an area has a persistently high level of unemployment, individuals out of work may be willing to work for substantially less than the prevailing wage. White (1987) has shown that unemployed workers often will be willing to accept significantly less than the prevailing wage in their previous employment. The longer their unemployment, the lower their reservation wage. Hence firms that consider only the prevailing wage may ignore opportunities to hire workers at reduced wages. Some corporate officials, however, may be concerned about future labor problems or pressures to increase wages in the future if they initially pay below the prevailing wage.

Productivity. Productivity is another factor that drives a wedge between the wage rate and unit labor costs. Productivity reflects output per worker. A highly skilled work force may be productive enough to have low unit labor costs even if a region's wage rate is higher than elsewhere. If the enhanced productivity is a result of superior local management rather than

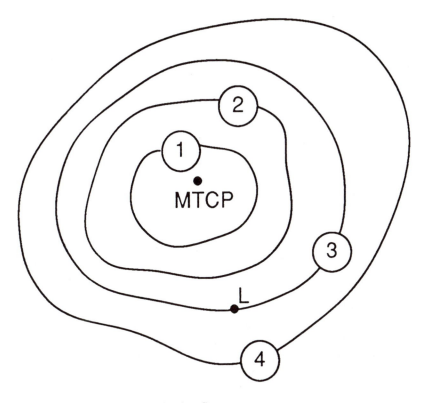

Figure 1.4. Variations in Production Costs

a more capable work force, however, the company may be better off locating in a low-wage region and applying their productivity enhancing management techniques at that location.

The labor environment. Poor labor relations may also reduce labor productivity. For instance, if a company is plagued with strikes, labor sabotage, or a work slowdown, productivity will be adversely affected. Fear of poor labor relations may be another reason that firms are reluctant to pay below the prevailing wage in an area even if they could hire workers at below average compensation.

In addition to the wage rate, state policies toward unemployment compensation, payroll taxes, and other costs associated with labor may affect labor costs. For example, a number of studies have indicated that it is not the level of state and local regulations that concern businesses as it is uncertainty over future regulatory policies (Corina et al., 1978).

Unionization may affect location through its perceived effect on current wages, future wages, or productivity. In light of several empirical studies, Schmenner (1981: 7) concluded that "right to work laws . . . are often the most effective public policies" for attracting new firms. Bartik (1984: 19) found that "a 10 percent increase in the percentage unionization of a state's labor force is estimated to cause a 30-45 percent decrease in the number of new branch plants."

Industrial filtering. The process of industrial filtering refers to the tendency of establishments to locate in metropolitan areas when they are new. Later in the product life cycle, after the kinks have been removed and the production process has become routine, establishments frequently relocate to smaller, rural communities. Differences in labor costs are the primary explanation for the industrial filtering model.

In the initial stages of product development and production, the higher skilled, higher cost labor associated with metropolitan locations is relatively important. In the earlier stages of development, firm dependence on external marketing and research experts, for instance, may be important. If the production process breaks down or if slight changes in the product are necessary, skilled engineers and other experts will be available in metropolitan regions to help make the needed changes efficiently. Later in a product's life cycle, as production becomes more routine, access to skilled labor becomes less important. Consequently, as firms mature, they will tend to relocate to areas where labor costs are lower.

High-tech firms are an important exception to the filtering process. They continuously operate at the earlier stages of their product life cycle. High-tech firms remain young as they progress from one new technology to another. These firms tend to cluster in regions that offer an abundance of technicians, engineers, and R&D research professionals (Premus, 1982). The Silicon Valley in California and Route 128 in Boston, Massachusetts, are world-renowned examples of major centers of high-tech activities in the United States.

Taxes and Business Climate

For many businesses, taxes are a small portion of total business costs. Consequently, many analysts have suggested that taxes are not an important locational factor. For instance, Carlton (1979) in an examination of the births of branch plant locations in three specialized industries concluded: "The evidence does not provide strong support for the proposition that taxes are a major deterrent to new business activity" (p. 15) "and the evidence provides little support for the proposition that state policies which improve business climate will stimulate new births" (p. 20). Carlton's findings were confirmed in a later study using similar data but a slightly different statistical model (Carlton, 1983).

On location surveys, business leaders have reported that tax factors are an important consideration. Some observers are justifiably skeptical of such statements. Business executives may simply respond so as to influence future tax policy. Accordingly, if business leaders say that they are likely to locate in low-tax areas, they may encourage public officials to reduce tax rates.

There is recent empirical evidence suggesting that taxes affect the location choices of particular types of businesses, at least in a small way. For instance, Wasylenko (1984) found that high personal income tax rates have a detrimental effect on economic growth. The logic of his findings suggests that plant managers may resist locating in high personal income tax states not so much because it will reduce corporate profits (personal tax rates do not directly affect corporate profits) but because the after-tax income of the managers will be reduced. Moreover, the business executives may perceive that the high personal tax rates may ultimately translate into pressures to offer higher wages to key personnel to offset the income tax burden.

Amenities and Quality of Life

Recently, many observers have suggested that quality of life directly contributes to economic development success. As discussed, as transportation costs fall and the production of information becomes increasingly important, firms become more "footloose." Accordingly, business decision makers may have more leeway in determining where they want to locate their businesses.

Amenities may boost economic development prospects in another way. Labor market theory suggests that workers may be equally happy living in an amenity-rich area and receiving lower wages or living in an amenity-poor area and receiving higher compensation. In other words, at the margin, workers may be willing to trade off higher wages for a more amenity-rich environment. Empirical studies have quantified the trade-off by estimating the value placed on a variety of amenities (Blomquist et al., 1988). Econometric studies have shown that the wage level for workers of comparable skill and experience may be significant enough to encourage some firms to locate in amenity-rich areas due to their lower labor costs. An explanation of the rise of the Sun Belt economies in Florida and California has been predicated on this cost-compensated wage theory.

Government

In the early 1800s, local governments were actively subsidizing canals and railroads to improve their regions' access to markets and materials (Benjamin, 1984). In recent years, however—particularly since the 1970s—government's role in locational choices has become much more pervasive.

The importance of taxes has already been discussed, but the role of government goes well beyond tax policies. First, government, particularly the federal government, plays a direct role in determining the location of major governmental facilities such as space centers, research laboratories, other scientific facilities, military installations, and administrative offices. The government's location decisions often create a surrounding economic environment that attracts additional private sector activity to the region.

Indirect influences. The role of government in the economy may also indirectly influence the location of economic activities. For instance, when USAir acquired Peidmont, the merger raised a number of important anti-trust questions. Senator Metzenbaum (D-Ohio) held hearings to determine whether antitrust legislation could be used to prevent the buy-out. During the hearings, the senator extracted statements from USAir officials that they had no plans to reduce their operations in Dayton. Although they did close their hub several years later, the influence of a powerful senator may have helped maintain the USAir hub in Dayton for a longer period than would otherwise have been the case. If it was not for serious excess capacity problems within the airline industry, the political actions could have resulted in the maintenance of the USAir hub in Dayton for a longer period.

Also, some firms that do significant work for the government may choose to locate a facility in a state or congressional district that is represented by a legislator with particular influence over that company's business. For instance, a defense contractor may find it useful to locate a facility in the district of a member of the Armed Services Committee. Thus, if the contractor needs help in dealing with the defense bureaucracy, it might be advantageous to be a constituent of a member of Congress with influence among Defense Department officials.

Political climate and international comparisons. Political climate is a regional locational factor that is generally the same in regions throughout a nation, particularly in the Western industrialized countries. As the world economy becomes more integrated, however, a larger number of companies are considering location sites throughout the world. One of the most important considerations of foreign investors is whether the government is stable and whether the political climate is compatible with good investments.

The political stability in the United States, Canada, and many other countries has contributed to an inflow of foreign investment in recent years.

Intergovernmental competition for jobs. State and local governments have created numerous programs and incentive packages in an attempt to attract business establishments to their areas. Frequently used incentives include infrastructure assistance, industrial revenue bonds, and tax abatement. Econometric studies have not been able to confirm the degree of effectiveness, if any, of specific subsidies or locational incentives. Nevertheless, 71% of new plants received at least one type of governmental assistance, according to a study by Schmenner (1981: 6). Apparently, many local economic development officials believe that their communities must offer a variety of locational incentives simply to keep up with incentives being offered elsewhere.

One of the criticisms of local development efforts is that cities compete without increasing the total number of jobs. One community gains only at the expense of another area. Zero sum games are particularly likely to occur when communities within the same metropolitan area compete for jobs. Concern over the zero sum game aspect of local incentives was expressed in the *President's National Urban Policy Report* (U.S. Department of Housing and Urban Development, 1988): "Place specific economic development policies often do nothing more than tax one place to improve conditions in another. The wealth of both places is not greater and may actually be less than it might have been" (p. II-2).

Although the possibility of a zero sum outcome exists, local incentive programs may provide incentives to create net new jobs. This would be the case if the incentives made it possible for the recipient firms to operate profitably or by building a set of community resources that encourage growth of existing businesses and the establishment of new businesses.

In addition to the possibility of a zero sum outcome, the use of local incentives has been criticized as being inefficient because the value of a subsidy to the recipient firm is often less than the cost to the taxpayers. For instance, new infrastructure to attract a firm may cost $5 million dollars to construct but the firms that are intended to benefit may value the infrastructure at only $4 million dollars.

Furthermore, local policymakers may be in an inferior bargaining position when negotiating with business executives over the location of a facility. One factor, called "information asymmetry," refers to the tendency for business representatives to have more information than local public officials. Another factor that may tilt the bargaining process in favor of businesses is the tendency of political officials to use the attraction of a new employer to shore up political support. A combination of these factors may explain why industrial location subsidy offers have become more generous during the slow growth period of the 1970s and 1980s.

Inertia

Inertia—the tendency to stay put—may be one of the most important locational factors; yet, it has not been intensively analyzed by locational theorists (Blair, 1991: 21). One reason for the neglect of inertia is that economists have focused on relocations and newly created facilities. Once a firm is established at a location, many forces operate to keep it there. When expansions are anticipated, they tend to occur in proximity to the original location. In fact, an expansion in place may not be viewed as a location choice by some specialists. Even when a relocation occurs, the firm is likely to choose a location near the current site because of what we call inertia. The term *inertia* does not necessarily imply, however, that a conscious decision to remain at a location has not been made.

One major aspect of inertia is that many businesses consciously prefer to maintain their labor force intact when they relocate or expand. Therefore, management may be reluctant to relocate a research facility to a new region fearing that some of the key researchers will choose to remain where they may have established social ties. Manufacturing or service-oriented companies may likewise wish to avoid the risk of losing key personnel from a major relocation. Schmenner (1982: 91) reported that more than 60% of plant expansions (as opposed to relocations) were motivated at least partly by the desire to maintain the management team.

Another reason for inertia is that firms tend to build relationships with other firms in the same area. Other local firms tend to be customers and suppliers. While potential customers and suppliers may be available at other locations, the relations already built at the existing locations may have value and may not be easily replicable at other locations. For instance, a market research firm may be reluctant to relocate from Nevada to Los Angeles because the increased distance may make it more difficult to deal with established customers and suppliers. Perhaps face-to-face meetings are few, but useful discussions at social occasions would be reduced due to the relocation. Good social relationships often make business dealings smoother and face-to-face meetings usually impart more information than telephone conversations. Similarly, supplier relationships may deteriorate with distance. Perhaps a survey research firm or an office services firm will be less accommodating at a new, distant location.

The realization that most new local jobs are created by businesses that already exist in a locality, coupled with Birch's (1979) well-publicized (but controversial) finding that existing small establishments have a high job creation potential, has brought about a major change in local economic development policies and practices. (More will be said about this in Chapter 13.) The efforts of local economic development practitioners have shifted from a focus on attracting large, generally manufacturing, firms from out of the area—pejoratively referred to as "smokestack chasing"—to seeking to encourage local firms to expand. Accordingly, local economic development agencies have established business assistance departments that assess local business needs, assist local firms in marketing, provide guidance regarding regulations that might hinder a firm's operations, and offer other services.

Agglomeration Economies and Coevolutionary Development

Agglomeration economies refer to benefits that accrue when firms locate in proximity to one another. Isard (1975: 113) stressed the importance of agglomeration economies in location theory: "An understanding of the development of cities and regions cannot be acquired without a full appreciation of the forces of agglomeration."

There are several types of agglomeration economies ranging from specific to diffuse. Some agglomeration economies reflect the cost advantages of consolidating operations of a firm at one facility by locating a cluster of plants near each other. Firms that buy and sell with each other may benefit from less specific agglomeration economies through reduced transportation and negotiating costs. Agglomeration economies can also occur among clusters of related firms. The advantages of a cluster of firms locating together include an enhanced labor force, sharing of specialized machinery, ease of attracting comparison shoppers, and an improved ability to imitate competitors.

Urbanization economies are the most diffuse types of agglomeration economies. They are benefits that accrue to a wide variety of firms when the overall level of economic activity in an entire metropolitan area increases. According to Mills and Hamilton (1984: 18), "The most important of such agglomeration economies is statistical in nature and is an application of the law of large numbers. Sales of outputs and purchases of inputs fluctuate . . . for random, seasonal, and cyclical and secular reasons." Firms in large urban areas will be better able to adjust to these fluctuations.

Agglomeration economies are an important locational factor in their own right, but they take on added importance when seen in the context of coevolutionary development. Once a center of agglomeration has formed, it becomes difficult for an establishment to move outside of the area of agglomeration because of the business dependencies that have evolved.

Many agglomeration economies develop through the process of coevolutionary development. Norgaard (1984) developed the ideal of "coevolutionary" development to describe how institutions in a locality develop symbiotic relationships. Norgaard's analysis places development in an evolutionary context. Institutions in the community evolve in ways that support the firm and the firm fills useful niches in the community. One implication of this perspective is that the forces of inertia will be stronger the longer a firm has remained at a particular site, making relocation more costly.

A Case Study

Many of the principles of location discussed in this review are illustrated in a case study of Diconix—an Eastman Kodak company specializing in advanced technology printing. It located in a major new facility in the Miami Valley Research Park in Dayton, Ohio, in 1984. The location of Diconix illustrates the importance of coevolutionary development and inertia, particularly as these forces operate in locational decisions of knowledge-based establishments. The case study also illustrates the importance of technological linkages, corporate decision-making processes, a supportive role for government, and the diminishing relative importance of several traditionally important locational factors.

Background

Diconix had its genesis in the Mead Corporation, a *Fortune* 500 firm with its world headquarters in Dayton, Ohio. As part of their corporate diversification program, Mead created the Advanced Systems Group to spearhead a wide variety of technologies related to paper products. Part of the Advanced Systems Group was a small division working with ink jet technologies, Mead Digital Systems (MDS). The technologies of MDS ranged from ink chemistry to machine design. The movement from paper products to printing is representative of the process Jane Jacobs described as "adding new work to old."

In mid-1982 Mead decided to sell MDS, which was not a major cash generator because of its heavy research investment. At the time, Mead desired an infusion of cash. MDS was scattered at three sites, two in the Dayton region and one in Texas, so its attachment to Mead did not involve a substantial space commitment.

Eastman Kodak, an integrated photo processing company, had an Information Systems Group that made copy machines. The company's corporate strategy included movement into printing. MDS appeared to be a good vehicle for expanding their printing capabilities. In mid-1983 Eastman Kodak purchased MDS and renamed it Diconix. The primary asset of MDS was the ink jet technologies under development and embodied in the personnel.

The physical location of Diconix was unsatisfactory. Even when the organization was within Mead, thought had been given to consolidation. One of the Diconix executives pointed to the need for synergy among the research, administration, and marketing staffs as a driving force in the decision to consolidate. Thus there was a need to find a new location for the newly formed Kodak subsidiary.

To understand the development of the Miami Valley Research Park where the facility eventually located, it is useful to understand the economic base of the region. The Dayton MSA is a region with one foot planted in traditional manufacturing operations and the other in advanced technology. (Dayton officials prefer the term *advanced technology* rather than *high tech* to indicate the region's comparative advantage in technologies near the production stage as opposed to basic scientific research.)

The strength of the manufacturing sector is concentrated in automobile production. In the advanced technology sector, the region is the site of a major Air Force aerospace research and logistics facility. Numerous small consulting firms that specialize in information system and aerospace technology support the Air Force base. Dayton is also the corporate headquarters of NCR, a prominent computer company (recently acquired by AT&T as part of a corporate strategy to link computers and telecommunication). Efforts by local business and government leaders to strengthen the advanced technology sector account for the development of the Miami Valley Research Park in the Dayton area. These efforts form part of the backdrop for this case.

A group of prominent local leaders worked with state government officials to create a research park, following the successful example of several other communities. Although local officials were following the example of others, they were aware of the importance of establishing an advance technology agglomeration focus. The research park concept was modified to account for its excellent location relative to national markets as well as the region's existing agglomeration economies. Accordingly, the Miami Valley Research Park accommodated certain light manufacturing and other activities that were not strictly research.

The organizers of the Miami Valley Research Park were successful in persuading the State of Ohio to contribute land that had previously been a state-run mental health institution to the park. The State of Ohio also

contributed funds for infrastructure improvements including new road-
ways, lighting, and a water-sewer system.

The Decision

After considering possible alternatives, Eastman Kodak made the deci-
sion to locate Diconix in a new building in the Miami Valley Research
Park. The principal decision maker was David Lehman, president of
Diconix. In discussing the locational decision, Lehman appeared quite
knowledgeable about what locational factors were important to high-tech
establishments and why those factors were important. The location of
Diconix can be divided into two decisions: the choice of locating in the
Dayton region and the decision to locate at that particular site.

Lehman reported that the main reason for locating in Dayton region was
the maintenance of the staff in a knowledge-based industry: "If I had a
totally new start-up situation, I might not have selected Dayton. However
I wanted to keep the team of technical experts together. This is a business
where our assets walk out the door at 5:00 p.m."

In fact, top Eastman Kodak executives expressed some interest in
locating Diconix in Rochester, New York, Eastman Kodak's world headquarters. Lehman resisted the Rochester location for two reasons. First,
the previously discussed desire to maintain the existing personnel base was
an overriding consideration. Second, there was a concern that Diconix would
lose its separate identity in Rochester, a city with 50,000 Eastman Kodak
employees. (The consideration of Rochester, nevertheless, indicates the attraction potential of a corporate headquarters for ancillary operations.)

A second factor in the selection of Dayton was the linkages that carried
over from the MDS group with people and institutions in the local economy.
Linkages with the former "parent," Mead, were important. In addition, Lehman
noted associations with three area universities and other corporations.

A final advantage of the Dayton region was that Dayton had "an atmosphere that the type of people we wanted to attract would like." Specifically, Lehman mentioned a warm, "family atmosphere" and a region that
was not too big. Lehman was concerned, however, that the region lacked
some cultural, entertainment, and leisure activities. After the location,
Diconix executives made efforts to strengthen the people-attracting infrastructure by leading an effort to revitalize downtown arts facilities.

When asked specifically about access to markets, transportation costs,
access to material inputs and labor, all of the prominent individuals
interviewed reported that, except for labor, traditional locational factors
were not important. Furthermore, labor was not important in the way it
traditionally has been. Diconix officials believed they could hire skilled

and semiskilled production employees with nearly equal ease at most metropolitan locations. They hired their engineering and scientific staff in a national market. Their primary labor concern was that the region be attractive to potential employees.

Several factors contributed to the selection of the Miami Valley Research Park site. First, the director of the nonprofit board that operated the park was James McSwiney, chairman of Mead Corporation. Ties between McSwiney and representatives of Diconix were probably an important, although intangible, factor in favor of the research park. Most of the Diconix employees, including Diconix President David Lehman, were with Mead prior to its purchase by Eastman Kodak.

Second, Diconix management was concerned that the building make a "statement" to the community and to employees. They also wanted a new facility designed to maximize communication among employees. In terms of architecture, the facility was designed to maximize communication by use of an open space design. Only through a new facility in a research park setting could these objectives be attained.

Finally, Ohio's contribution in the form of land and infrastructure kept site costs attractive, although no additional state subsidies were given. The state contributed about $10 million for infrastructure and land improvements.

Postscript

By 1988 Diconix had grown to about 840 employees, more than twice the start-up number. Eastman Kodak underwent a major restructuring during that year, however, as did many major corporations. As a result, several units within Diconix were transferred to Rochester and employment fell to about 350. The name was also changed. The Dayton facility became part of the Copy Products Divisions and a degree of autonomy was lost.

Since the 1988 reorganization, other name changes have occurred as responsibility for the facility has been shifted among different corporate offices, although the basic operations of the Eastman Kodak unit has not changed significantly. Currently, the facility is part of Eastman Technology Ventures. As the focus of their efforts have been modified, new linkages with Dayton area consultants and suppliers have been formed.

A Critical Appraisal

Location theory is one of the most developed branches of urban and regional economics. The principles described in this chapter help explain not only the location of individual establishments but also how locational decisions combine to shape metropolitan systems. Location theory has proven to be *a valuable tool kit* both for business decision makers seeking to make better location decisions and for *public officials seeking to make their communities more attractive sites.* Nevertheless, there are several areas where location theory could be improved.

Motives

Location theory is normally predicated upon the assumption that decision makers are motivated to maximize expected profits. This assumption, of course, is typical of most economic models.

Isard (1975: chap. 4) suggested that business officials may frequently select locations that will not maximize expected profits. Executives who prefer a safe location may be willing to sacrifice some profits for the security of making a decision that is safe from criticism or that will have a high probability of earning at least a minimum threshold profit level. The search for security at the expense of some profits is particularly likely among companies where executives own only a small part of the company. In this case, management may be less willing to pursue riskier strategies to raise profits than the shareholders may wish.

Recent analysis of business decisions suggests that many executives may be more concerned with short-run outcomes than maximizing profits over the long term. If executives believe their performance will be judged over a two- or three-year time period (or less in some cases), locational choices will be biased toward the short term.

The desire for security at the possible expense of profits may also influence the advice given by consultants. Consultants have an incentive to suggest locations that have good reputations over potentially profitable sites that are not recognized. For instance, a locational consultant might feel more comfortable suggesting that a high-tech client select a site in the Silicon Valley rather than Canton, Ohio, even if the Canton site is objectively equal to the Silicon Valley location.

Hoover (1948) argued that locational theorists need not be concerned about motives of decision makers. In the long run, establishments will be forced to choose the most profitable location or they will go out of business:

> A good analogy is the scattering of certain types of seeds by the wind. These seeds may be carried for miles before finally coming to rest, and nothing makes them select spots particularly favorable for germination. . . . Because of the survival of those which happen to be well located, the resulting

distribution of such from generation to generation follows closely the distribution of favorable growing conditions.

The analogy is brilliant and has some validity, but it is too sanguine about the ability of market forces to rectify bad decisions. The economy is not so competitive that nonoptimal locations will be quickly eliminated. In fact, bad locational choices could continue for decades. In the meantime, other institutions could develop that support the initially suboptimal location. As the process of coevolutionary development continues, an initially suboptimal location could become a more suitable environment.

Public Decisions

Our understanding of the motives of government officials is also weak. In light of the importance of the public sector, public location decisions need to be better understood. Although location theorists can identify an optimal site for a fire or police station, there is difficulty in explaining actual decisions. Public officials are motivated by factors other than efficiency just as are private sector decision makers.

Many locational choices are the result of bargaining between state or local governments on the one hand and businesses on the other. The extension of location theory to incorporate bargaining between businesses and several units of governments would enhance the realism of locational models.

Predictability

One mark of a good theory is predictability. Ideally, location theory should be able to predict where a facility will locate. In reality, the predictability in industrial location theory is poor. Moreover, as the list of significant locational factors has lengthened, prediction has become less accurate.

Many areas have very similar locational characteristics. The differences between the regions are particularly small when compared with the ability of location specialists to accurately quantify the theoretically important concepts such as labor costs, quality of life, and so forth. The increased importance of state and local incentives has increased the difficulty of prediction.

Spatial Ambiguities and Locational Benefits

Spatially, it is sometimes unclear where the benefits of some locational factors begin and when distance renders these factors no longer important.

For instance, if an establishment is located within a particular tax jurisdiction, it will receive the appropriate tax benefits, so there are no spatial ambiguities. Most locational attributes are subject to a "distance decay" function, however. It seems clear that the advantages of being near universities, railroad stations, customers, and amenities decrease at a certain distance from the attribute. The benefits do not immediately disappear, however. They probably decline either quickly or slowly until the benefits become negligible. Location theory would be improved if economists had improved knowledge as to how the advantages and disadvantages of locational attributes are influenced by distance.

When Is an Establishment Ripe for Change?

Historically, location theory focused on the location of new establishments. The relocation decision has received less attention. Firms make locational decisions only at key times in the business life cycle of their companies. Examples of locational decision points include when an existing plant is operating at capacity, a merger has occurred, a change in product line is anticipated, or the physical plant is near obsolescence.

Partial Equilibrium Framework

Finally, most models assume that the location of the firm will not affect the environment in any relevant way. For instance, it is assumed that the location choice will not affect the wage rate, the quality of life, the size of markets, or other factors.

The assumption that a location will not affect the environment is appropriate for small company locational decisions but may be inappropriate for large establishments. For instance, the location of the GM Saturn plant in Spring Hill, Tennessee, changed the labor, cultural, and transportation environment immediately and set in motion forces for continuing changes in the region. Similarly, when a large retail store locates in an area, transportation and shopping patterns may be affected. These changes could, in turn, affect the success of businesses at the location. The development of a better understanding of how locational choices affect the environment is an important area for future research.

Conclusion

Location theory has its origins in the traditional economic paradigm of profit maximization. The traditional model worked well when a few easily qualifiable cost factors dominated the location decision. Analysis of the basic cost factors continues to provide important insights. Traditional cost factors have diminished in relative importance, however, while the list of relevant variables has expanded to include many other factors, some of which are difficult to quantify. Consequently, deterministic, profit maximizing models of locations have become more complex, unwieldy, and less reliable.

Location theory has been extended beyond static profit maximizing models. At the frontiers of location theory, analysts are seeking insights into motives, the role of the public sector, and the interaction between a facility and the host environment.

References

Bartik, T. (1984). "Business Locational Decisions in the U.S.: Estimates of the Effects of Unionization, Taxes, and Other Characteristics of States." *Journal of Business and Economic Statistics* 3: 14-22.

Benjamin, Robert Cook (1984). "From Waterways to Water Fronts: Public Investment for Cities, 1815-1980" in R. Bingham and J. Blair (eds.), *Urban Economic Development* (pp. 23-46). Beverly Hills, CA: Sage.

Birch, David (1979). *The Job Generation Process*. Washington, DC: Government Printing Office.

Blair, John P. (1991). *Urban and Regional Economics*. Homewood, IL: Irwin.

Blair, John P. and Robert Premus (1987). "Major Factors in Industrial Location: A Review." *Economic Development Quarterly* 1: 72-85.

Blomquist, Glenn C., Mark C. Berger, and John P. Hoehm (1988). "New Estimates of Quality of Life in Urban Areas." *American Economic Review* (March): 89-107.

Carlton, D. (1979). "Why New Firms Locate Where They Do: An Econometric Model" in W. Wheaton (ed.), *Interregional Movements and Regional Growth* (Coupe Paper 2; pp. 15-16). Washington, DC: Urban Institute.

Carlton, D. (1983). "The Location and Employment Choices of New Firms: An Econometric Model with Discrete and Continuous Endogenous Variables." *The Review of Economics and Statistics* 65: 440-449.

Corina, G., W. Testa, and F. Stocker (1978). "State and Local Incentives and Economic Development" (Urban and Regional Economic Development Series, No. 4). Columbus, OH: Academy for Contemporary Problems.

Hoover, Edgar M. (1948). *The Location of Economic Activity*. New York: McGraw-Hill.

Isard, Walter (1975). *Introduction to Regional Science*. Englewood Cliffs, NJ: Prentice-Hall.

Mills, Edwin S. and Bruce W. Hamilton (1984). *Urban Economics* (3rd ed.). Glenview, IL: Scott Foresman.

Norgaard, Richard (1984). "Co-evolutionary Development Potential." *Land Economics* 60: 159-167.

Premus, R. (1982). *Locational High Technology Firms and Regional Economic Development* (Joint Economic Committee Print). Washington, DC: U.S. Congress.

Schmenner, Rodger W. (1981). "Location Decisions of Large Firms: Implications for Public Policy." *Commentary* (January): 3-7.

Schmenner, Rodger (1982). *Making Business Locational Decisions.* Englewood Cliffs, NJ: Prentice-Hall.

Stevens, Benjamin H. (1985). "Location of Economic Activity: The JRS Contribution to the Recent Literature." *Journal of Regional Science* 25 (4): 678.

Thisse, Jacques Francois (1987). "Location Theory, Regional Science, and Economics." *Journal of Regional Science* 27 (4): 519-528.

U.S. Department of Housing and Urban Development (1988). *The President's National Urban Policy Report.* Washington, DC: Author.

Wasylenko, Michael (1984). "The Effects of Business Climate on Employment Growth in the States Between 1973 and 1980" (Report for the Minnesota Tax Study Commission).

White, Sammis B. (1987). "Reservation Wages: Your Community May Be Competitive." *Economic Development Quarterly* 1 (February): 18-29.

Suggested Readings

Blair, John P. (1991). *Urban and Regional Economics* (Chaps. 2-4). Homewood, IL: Irwin.

Blair, John P. and Robert Premus (1987). "Major Factors in Industrial Location: A Review." *Economic Development Quarterly* 1 (February): 72-85.

Charney, A. H. (1983). "Intraurban Manufacturing Locational Decisions and Local Tax Differentials." *Journal of Urban Economics* 14: 184-205.

Hoover, Edgar and Frank Giarratani (1984). *An Introduction to Regional Economics* (3rd ed.). New York: Knopf.

Norgaard, Richard (1984). "Coevolutionary Development Potential." *Land Economics* 60: 159-167.

Premus, Robert (1982). "Location of High-Tech Firms and Regional Development." Washington, DC: U.S. Congress, Joint Economic Committee.

Rosen, S. (1974). "Hedonic Markets and Implicit Prices: Product Differentiation in Pure Competition." *Journal of Political Economy* 82: 34-55.

Schmenner, Rodger (1982). *Making Business Locational Decisions.* Englewood Cliffs, NJ: Prentice-Hall.

Skoro, Charles L. (1988). "Ranking of State Business Climates." *Economic Development Quarterly* 2 (May): 138-152.

Thisse, Jacques-Francis (1987). "Location Theory." *Journal of Regional Science* 27 (November): 519-528.

White, Sammis B. (1987). "Reservation Wages: Your Community May Be Competitive." *Economic Development Quarterly* 1 (February): 18-29.

2

Theories of Regional Development

ARTHUR C. NELSON

This chapter reviews the major theories describing regional development, identifies the apparent theoretical underpinnings of U.S. regional development policy, and poses the theoretical underpinnings of future regional developments given current policies and trends. The focus is on applying regional development theory to local development policy.

But, first, what is meant by *regional development?* The term refers to change in regional productivity as measured by population, employment, income, and manufacture value added. It also means social development such as the quality of public health and welfare, environmental quality, and creativity.

For whom is regional development beneficial? European settlement of North and South America was attributable to expansionary policies of European institutions. Colonies served mother countries through exploitation of resources. To mother countries, regional development meant creating greater wealth. To native Americans, regional development meant genocide. To colonialists, regional development meant the hope of a higher quality of life relative to their social status in the mother country. Ultimately, colonists saw greater opportunity for wealth and influence by seceding from the mother country to form their own countries and institutions. These new countries extended regional development further into the hinterland. Now most of the New World, to a large extent, has been developed.

Within the New World, a number of different tensions arose. Dominant regions exploited the hinterlands. In the United States, the tension between the industrial North and the agrarian South led to the Civil War. In more recent times, differences in regional development are viewed as attributable to some regions being chronically underdeveloped, such as Appalachia, or undergoing structural change associated with business and technological cycles, such as the manufacturing belt. Effective regional development policies will be different in both regions yet federal regional development policies often do not differentiate between regions.

In this chapter, *regional development* does not mean the exploitation of virgin territory or the domination of one culture over another but, instead, improving the conditions of chronically underdeveloped regions or regions undergoing cyclical change.

A note on important terms is in order. The regional development literature distinguishes between *core* (or *center*) and *periphery*, *growth center* (or *pole*) and *hinterland*, and *leading* and *lagging* regions. The problem for policymakers and practitioners is knowing whether their region is core or periphery, pole or hinterland, leading or lagging. Different development policies may come to play in each situation.

The "core (or center) and periphery" concept distinguishes between regions on a global scale—regions can be composed of entire nations or collections of states. In colonial times, England and Spain were the core (or center) while the colonies of North and South America were the periphery. In modern times, at least until after World War II, the northern states were considered the core (or center) while the southern states and much of the West were considered the periphery.

The "growth center (or pole) and hinterland" concept distinguishes within a regional context. Growth centers are urban or extended metropolitan areas, here called "urban fields"; hinterlands are outside the urban field. Thus, within both core and periphery regions, there will be growth centers surrounded by hinterlands.

The "leading and lagging" concept distinguishes advanced regions from underdeveloped regions at both the global and the regional levels. The periphery can be composed of growth poles that are leading regions while their hinterlands are lagging. The core can have hinterlands that are leading regions.

Theories of Regional Development

There are two dominant schools of thought on regional development. The *development-from-above* school views regional development as essentially emanating from the core and growth centers and trickling out to the periphery and hinterlands. In its crudest sense, this school views regional development as starting from worldwide demand or critical innovation that filters down to national, subnational, urban units, and hinterland units. The *development-from-below school* does not necessarily dispute the path of development-from-above but, instead, argues for regions to take control of their own institutions to create the life-style desired in the region.

Both schools have as a common ground the application of economic base theory. Economic base theory presumes that regional development is driven by economic rules applied to space. It focuses on firm productivity, firm location, and agglomeration economies leading to cities. It explains how systems of cities are formed and describes the economic relationships between cities within a hierarchy. Regional development is sustained through vertical and complementary linkages among industries. Three kinds of external relationships are critical for regional development: trade characterized as imports and exports of goods and services; migration of people in their capacities as both consumers and workers; and migration of other factors, principally capital for investment.

Economic activities of a region can be divided between industries producing goods or services for export to other regions and industries producing goods or services for local consumption. The economic development of a region depends on its ability to raise the volume of exports relative to consumption of locally produced goods and services. The ability of a region to sustain long-run economic development depends on its ability to continue to export goods and services. This requires attracting capital and appropriately skilled labor essential to sustaining its development (North, 1956).

In many regions, however, there is much to be gained by producing goods and services for the local population that a region must otherwise import. Where large regions can replace goods and services that are imported with locally produced goods and services, the flow of capital to other regions is reduced. Thus residentiary services—which are population-serving enterprises such as hair dressers, family doctors, and auto mechanics—and industries that produce goods for local consumption—such as processed dairy goods—are necessary to sustain development in any region (Tiebout, 1956). These principles apply to both urban places and systems of cities.

In addition to similar treatment of economic base, both schools of thought recognize that regions will progress through stages of growth. There are two general views of how regions evolve from lower stages to higher stages of development, one based on the work of Walt Rostow and the other on that of John Friedmann.

Rostow (1960) sees regional development progressing in five stages: traditional, takeoff preconditions, takeoff, maturity, and mass consumption. Progression from one stage to another may be delayed or rendered unachievable by a variety of factors. His model is meant to apply to nations but it is useful in the regional context as well.

A region in the traditional stage of development is one in which there is limited availability of technology relative to other regions and there probably exists a hierarchical social structure. This describes the South between the Civil War and World War I. It may still apply to some substate regions of the South.

In the second stage, the region's economic and social structure begin to change. This occurs when leading regions make investments in lagging regions in transportation, communication, and manufacturing processes for the purpose of exploiting natural resources. During this stage, managers and skilled labor are transferred to the lagging region. A new social and political elite emerges. One example is the change in economic, social, and political composition of the South attributable to the Tennessee Valley Authority.

Takeoff occurs when an external stimulus, such as a major war, infuses investment by leading regions into lagging regions, and when the new social and political order works to sustain that investment. In the southern context, World War II was the external stimulus. It led to investment into military-related operations such as major research and production centers. After the war, the research and production infrastructure was sustained in part through continued national investment and by inducing private sector investment in industries such as petrochemical, electronics, and aerospace.

Rostow suggests that the takeoff stage lasts 20 to 30 years. The South, for example, entered the maturing stage in the 1970s, about 25 years after World War II. Many formerly imported goods and services are produced locally. Miernyk (1977) confirms that, by 1975, the South's economy had diversified its economic base so much that the proportion of workers engaged in agriculture had fallen to the national average while employment in communication, transportation, utilities, and government grew to exceed national averages.

The fifth, mass consumption stage occurs when a region exports many goods and services that it formerly imported. The South may already be in this stage.

Friedmann (1966, 1967) offers a slightly different but enlightening view. Characterizing a center-periphery relationship, Friedmann suggests a continuum over time in which the relationship between the center and its periphery changes. In the preindustrial stage, there is very little interaction. Regional economies are colonial in nature and relate directly to the mother country. As colonial ties are severed, the regional economies find ways to relate to each other principally in the form of coinage and conventions on trade and tariff. With the rise of industrialization, some regions become exploited by others. At this stage, the center virtually lives off the periphery; the periphery develops only as a function of further exploitation by the center. Only in the postindustrial stage does the periphery have the opportunity (or the need) to develop economic independence.

Friedmann goes on to pose how regions relate to one another in a postindustrial economy. He divides a theoretical nation or superregion into five different kinds of individual regions. The "core" region has characteristics of his "urban field" concept (Friedmann & Miller, 1965). The "upward transitional" regions experience net in-migration and have need for new capital. "Downward transition" regions experience out-migration and may be overcapitalized or have obsolete economic structures. "Resource frontier" regions are devoted to natural resource exploitation such as mining, forestry, fishing, and agriculture. He also includes "special problem" regions as a catchall for otherwise undefinable regions. Each region has its own development opportunities and constraints.

Beyond the similarities in treatment of economic base and stages of development, the two schools of thought diverge. I will first examine the development-from-above school, then the development-from-below school, and then hazard a synthesis.

The Development-From-Above School

This school is rooted in traditional regional and neoclassical economic models. It assumes that regional development occurs when stimulated by exogenous forces such as export markets, investment from outside, and migration. Regional development is seen as a product of price equilibrium and disequilibrium. This school includes a political-economic view.

Price Equilibrium Models of Regional Development

Several models suggest that regional development occurs as a process of equilibration between regions. Many models aim to explain international development patterns but they are applicable to regions within a nation as well.

Heckscher (1919), Ohlin (1933), and Balassa (1961) develop the *price-equalization* model wherein reasonably mobile factors of production will seek locations generating the highest return. Investment by southern textile manufacturing firms in the Orient is an example of this. As investment in underdeveloped regions increases, however, so does the competition for production factors, resulting in increasing factor prices. Over time, returns to factor investments equalize. In the 1980s many textile operations were moved back from overseas to southern states as the price of labor, inputs, management, and distance rose.

The relocation of manufacturing from northern to southern and western states is at least partly explained by differential factor prices. Wheat (1976, 1986) explained that a high percentage of manufacturing job change between 1949 and 1967, and between 1963 and 1977, was related to differential factor prices such as labor and energy. He also noted that improvements in transportation reduced the cost of shipping goods from southern processors to northeastern markets.

Myrdal (1957) suggests a leading-lagging approach to explaining regional development. He develops the theory of *cumulative causation*. Leading regions possess initial comparative advantage due to location, infrastructure, and other factors. Agglomeration results in ever-increasing investment. Little investment moves from leading regions to lagging regions. Investment that does occur is controlled by leading region elites to assure economic dominance. Lagging regions are further inhibited in development because of backwash factors. Skilled workers, educated individuals, business leaders, and venture capital that may emerge in the lagging region will be drawn to the leading region to seek higher returns. Goods and services produced in the leading regions are sold to the lagging regions at such low prices that local industries cannot compete.

But one feature of leading regions is that they tend to spread out into lagging regions. Most lagging regions will have some comparative advantages, principally in natural resources, that result in positive investment flow. When the spread effects become stronger than the backwash effects, cumulative causation leads to development in the lagging region. Public policy can be used to induce investment in lagging regions. This may be done by discouraging certain kinds of investment in leading regions. The effect is price equilibration in the longer term.

Perroux (1955) and Hirschman (1958) develop the *growth pole* concept where development in hinterland regions is fueled by expanding metropolitan centers. In this context, investment trickles out from the growth pole or growth center to the hinterland. It may take the form of manufacturing investment (Nelson, 1990) or the settlement of population into the exurbs, which were formerly part of the hinterlands (Nelson and Dueker, 1990). Trickle effects can be seen within an urban field as a process of deconcentration from centers to hinterlands and from core to peripheral regions.

Richardson (1985) combines notions of cumulative causation and growth poles in his *autonomous growth centers* (AGC) model. One limitation of the growth pole model is that it is usually used to explain development patterns of an individual, often substate region. Richardson suggests a national system of AGCs where growth rates differ among AGCs over time based on agglomeration economies and locational advantages. He asserts that hinterland growth is highly correlated with AGC growth. Older AGCs suffer from weakened economies of agglomeration as new capital is diverted to younger AGCs. The diffusion in investment and population away from older to younger AGCs (and their hinterlands) starts gradually. In this respect, life-cycle theory explains regional development patterns over time. At some point, however, the diffusion compounds on itself and creates the dynamics of cumulative causation. On the other hand, Richardson's model is unsatisfactory in dealing with the phenomenon seen during the past 20 years of hinterlands among older AGCs growing while the AGCs themselves decline (Nelson, 1990).

Dynamic Disequilibrium

Schumpeter (1934, 1939, 1947) observes that one feature of capitalism is its propensity to destroy old regimes and create new ones. This is known as *dynamic disequilibrium*. Obsolete products and processes are replaced by more timely and efficient ones. The infrastructure of a leading region may succumb to aging and obsolescence. Investment in new industries may be more efficient in the lagging region. Thus manufacturing relocation from the Rust Belt to the Sun Belt is explained in part by the greater investment returns available in the South and West relative to tearing down and rebuilding infrastructure in the North.

Product life-cycle theory (Schumpeter, 1939; Vernon, 1966) expands on the concept of dynamic disequilibrium. A new product is developed and exported, resulting in rising regional incomes. The product is produced in other regions when there is sufficient local demand. Indeed, the innovating region may eventually import the very product it developed. The concept can be applied to regional development. Consider microelectronics production (versus research and development). Microelectronics technology was initially developed and mass produced in the San Francisco Bay area. But the Bay area has become a costly place for workers to live. When assembly processes achieved certain economies, high-tech firms realized that with minimal investment and work force training much of the assembly process could be done in the less costly lagging areas of California or other states. As new manufacturing centers emerge in the hinterland, however, new forms of innovation in assembly and even in production are possible (Glasmeier, 1990).

Weinstein et al. (1985) offer a *regional life-cycle* model built upon Nikolai Kondratieff's long waves (see also Hall, 1990). New enterprises emerge in lagging regions because leading regions are strapped with obsolete and unprofitable infrastructure. But regional life-cycle theory also holds that newly developed regions will themselves decline and by that time bypassed regions will have been retooled. One should thus expect an eventual decline of the Sun Belt and reemergence of the Rust Belt (Nelson et al., 1992; Rosenfeld, 1992).

A Political-Economic View of Development-From-Above

While lagging regions may grow and leading regions may decline, it is a particular group of individuals and institutions—the capitalist class—

that may cause some regions always to dominate and others always to be subservient. Richardson (1985) summarizes this view. First, cities dominate their hinterlands as they invest in the means to extract from the hinterlands resources that make cities richer. Rural savings are attracted to cities to earn higher yields. Second, spatial differentiation of production is not due to uneven distribution of natural resources or of different transportation cost functions but to the concentration of capital in large production systems. Third, because capital flows from cities only to those hinterland locations where returns are highest and not necessarily to where such investment would have the greatest social outcomes, there will always be parts of the hinterland that are deprived of investment. Fourth, because capital is mobile, it will flow to areas offering the greatest opportunities for exploitation such as areas with high nonunion membership.

Along the same lines, mesoeconomic power (Holland, 1976)—wielded by very large firms or, one might argue, investment pools such as pension funds—dominate the distribution of development among regions. The leading regions within which mesoeconomic interests operate grow ever more influential through ever-increasing control of industries and the economy. The result is that mesoeconomic firms locate in leading regions and the lagging regions are left with smaller scale, lower profit industries.

The Evidence and Assessment

As Markusen (1987) observes, the United States has pursued two sets of regional development policies each geared to different levels of space. At one level, it has pursued development of periphery regions such as the South and parts of the West. At another level, it has stimulated population and economic deconcentration from growth centers to hinterlands. In both cases, U.S. policy seems to be based on *cumulative causation* theory. Those policies directly affect the interaction between core and periphery and between the centers and the hinterlands. Bourne (1980) recounts how U.S. development policies favor undeveloped over developed areas, suburban and rural highways over urban mass transportation systems, new housing over rehabilitation, low-density over high-density housing, and new water and wastewater systems over rehabilitation of existing systems. The result is not that regional development becomes concentrated in growth centers but that growth centers themselves spread out into the regional landscape.

Although policies appear to be based on *cumulative causation* they were never intended to create autonomous growth centers. Whether intended or not, this policy orientation has integrated of growth centers with their hinterlands. Nelson and Dueker (1990) demonstrate that about 80% of the nation's population now live within a functionally integrated growth center-hinterland landscape larger than Friedmann and Miller's (1965) urban fields or Berry and Gillard's (1977) commuting sheds. They call this the "exurbanization" of America.

Some evidence on the success of development-from-above is offered here. It includes general summaries of the patterns of regional population and manufacturing employment growth and decline in the last quarter of the twentieth century. It also includes a review of the determinants of manufacturing shifts among regions.

Core-to-periphery manufacturing. Nineteenth-century firms had to be near the source of markets, labor, materials, and transportation facilities. Those factors strongly favored manufacturing location in New England, the Middle Atlantic, and the Midwest. In most of the twentieth century, however, population and associated economic activity has dispersed from the manufacturing belt (Cohen and Berry, 1975; Webber, 1984). Figure 2.1 shows one way in which to divide the nation into core (Northeast) and periphery (Midwest, South, and West) regions. Table 2.1 shows the changing share of manufacturing activities by census region since 1899.

General trends are clear. The Northeast saw the greatest loss of manufacturing jobs since World War II. Total manufacturing employment in the Midwest remained fairly constant but rose steadily in the South and in the West. The South has emerged as the nation's dominant manufacturing region. Indeed, virtually all of the nation's loss of manufacturing jobs since its peak in 1977 can be attributed to losses in the Northeast and to a lesser extent in the Midwest. Since 1967 the Northeast lost 1.2 million jobs, while the Midwest lost 700,000 jobs since 1977. The South and West combined added more than 2.2 million jobs since 1967. To the extent that U.S. regional development policy intended to elevate lagging regions' economies at least with respect to manufacturing, it has succeeded.

On the other hand, between 1982 and 1987, the manufacturing base of the Midwest expanded while that of the South held steady. The South added about 23,000 manufacturing jobs in this period while the Midwest added 102,000 jobs. These trends suggest that the cyclical nature of manufacturing investment could be swinging in the direction of the Midwest. The West added the greatest number of manufacturing jobs in this period—167,000—while the Northeast continued its two-decade decline by losing 442,000 jobs. One reason for the resurgence of midwestern manufacturing is its large base of skilled labor, physical infrastructure, level of public investment in education, and use of tax concessions to reduce taxes on capital (Nelson et al., 1992).

What are the particular determinants of core-to-periphery manufacturing shifts? Wheat (1986) estimates that the leading determinants of change

(text continued on page 38)

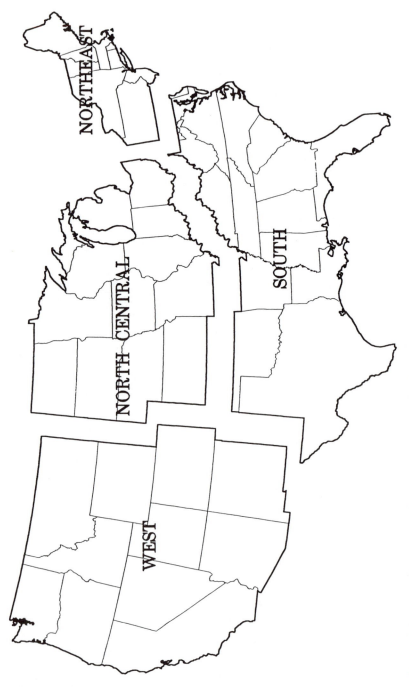

Figure 2.1. Census Regions and Divisions: Contiguous 48 U.S. States (county classifications: 1992)

36

TABLE 2.1. Employment in Manufacturing by Region: 1947-1987 (numbers of employees in thousands), Percentage of U.S. Total

Year	Nation	North-east	% of U.S. Total	Mid-west	% of U.S. Total	South	% of U.S. Total	West	% of U.S. Total
1987	18,950	4,357	23	5,508	29	5,839	31	3,246	17
1982	19,040	4,799	25	5,406	29	5,816	30	3,073	16
1977	19,590	5,008	26	6,272	32	5,593	29	2,717	14
1972	19,029	5,306	28	6,135	32	5,221	27	2,367	13
1967	18,492	5,573	30	6,059	33	4,553	25	2,305	12
1963	16,235	5,190	32	5,239	32	3,780	23	2,025	13
1958	15,423	5,242	34	5,107	33	3,382	22	1,779	12
1947	14,294	5,429	38	5,109	36	2,710	19	1,054	7

SOURCE: Data from 1987 Census of Manufacturing, U.S. Department of Commerce.

among states in manufacturing jobs, based on percentage change of manufacturing job growth between 1973 and 1977, were market accessibility at 55%, moderate climate at 15%, rural attraction at 11%, nonunionized and lower wage labor at 6%, thresholds at 5%, and retirement attraction-amenities at 4%. These factors combine to explain 96% of the variation in manufacturing job change.

In a study of the variations in the structure of metropolitan and nonmetropolitan manufacturing production, Blackley (1986) found that metropolitan counties possess the significant agglomeration economies needed for product development and process innovation. Statistically significant variables included value added by manufacturers, production worker hours, payroll, and state's manufacturing taxes, among other factors.

Center-to-hinterland manufacturing. While development has extended from the traditional core of the Northeast to the frontier and lagging regions of the Midwest, South, and West, what is happening between centers and their hinterlands? Recent research suggests that centers are becoming more integrated with their hinterlands.

In 1910 the U.S. suburban population was only 5.4 million. In the next three decades, suburban America experienced growth rates at twice the national average. Manufacturing activities followed, particularly those small-scale production firms that were market or labor dependent (Lessinger, 1986). The exodus from the central city reached new heights in the middle of the century. The National Industrial Conference Board (NICB) found that central cities lost 10% of their manufacturing jobs between 1947 and 1958. This was accompanied by a rise in manufacturing activities in suburban localities and in regions outside the manufacturing belt (Cohen and Berry, 1975).

An example of how centers are becoming integrated with their hinterlands is offered in my work (Nelson, 1990). I adapted Berry and Gillard's (1977) *commuting shed* definitions to devise exurban areas based on urban fields conceptualized by Friedmann and Miller (1965). I then investigated the distribution of population, total employment, and manufacturing employment among urban, suburban, exurban, and rural counties between 1965 and 1985. Figure 2.2 illustrates the national pattern of centers and their hinterlands.

Although total employment in large central counties (central counties of metropolitan areas with more than 500,000 population in 1985) rose by 51%, from 22.6 million to 34 million, manufacturing employment fell from 7.6 million to 7.1 million. In suburban counties (defined as noncentral metropolitan counties in 1960), employment rose by 111%, from 5.6 million to 11.9 million, but manufacturing employment rose by only 16%, from 2.3 million to 2.6 million. Employment in small central counties

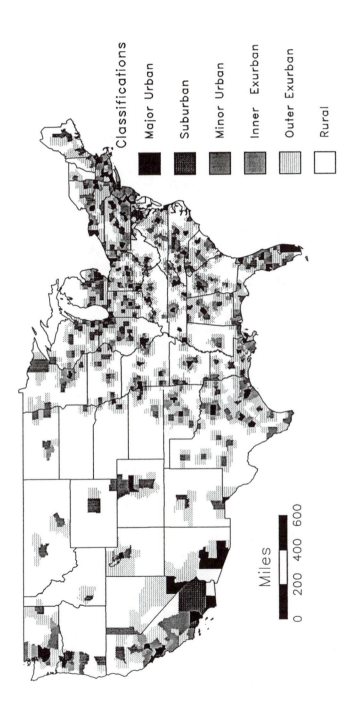

Figure 2.2. Center-Hinterland Regions of the United States

NOTE: The "center" is composed of city centers and suburbs. The hinterland is identified as "rural." The area of transition, indicating cumulative causation at work, is identified as "exurban." Map prepared by David S. Sawicki.

(defined as central counties of metropolitan areas with less than 500,000 population in 1985) rose from 8.2 million to 14.6 million, or 78%, while manufacturing employment rose from 3.2 million to 3.5 million, or 10%. Employment in rural counties rose from 3.2 million to 5.6 million, or 77%, while manufacturing employment rose from 1 million to 1.5 million, or 46%. (See Tables 2.2 and 2.3.)

I found that overall employment in exurban counties (located beyond suburban counties and up to 100 miles away from central cities) rose by 85%, from 7.6 million to 14.1 million. Most interesting is that manufacturing employment rose from 3.4 million to 4.5 million, or 31%. This was the largest increase in manufacturing jobs among all areas during this period. (See Tables 2.2 and 2.3.)

I also found that while total manufacturing employment in the contiguous 48 states rose from 17.5 million to 19.3 million—a change of 1.8 million—manufacturing employment outside central cities rose from 9.9 million to 12.1 million—a change of 2.2 million. Thus, while exurban areas accounted for only 20% of the total number of jobs created during this period, they accounted for 48% of the location of new manufacturing jobs. More important, they accounted for 61% of the entire share of manufacturing change. By comparison, suburban counties accounted for 17% of the total increase and 21% of the shift, small central counties accounted for 6% of the increase and 18% of the shift, and rural counties accounted for 22% of the increase and 27% of the shift. (See Tables 2.2 and 2.3.)

From 1960 to 1985 exurban counties added more population than any other area evaluated. Population rose from 42.5 million to 59 million, an increase of nearly 16.5 million. Exurban counties accounted for 30% of the share of continental U.S. growth. Exurban counties grew faster than all other counties in both real numbers and share of growth. By contrast, large central counties accounted for a quarter of the increase, suburban counties accounted for 18%, small central counties accounted for 21%, and rural counties accounted for less than 7%. (See Table 2.4.)

What are the determinants of the center-to-hinterland (or exurban) manufacturing shift? My colleagues and I (Nelson et al. 1992) used regression analysis to find that manufacturing growth in the exurbs is negatively associated with metropolitan manufacture value added, manufacturing employment as a percentage of total employment, union membership, property taxes, and distance to a small airport, and positively associated with the adult population percentage of high school graduates, expenditures in public education, and farm population. Within the exurbs, manufacturing growth is not associated with highway, large airport, or central city accessibility. The reason may be that, within any given exurban area, such factors are almost always present.

(text continued on page 44)

TABLE 2.2. Distribution of Total Employment by County Class, 48 Contiguous States: 1965-1985 (figures in thousands)

County Class	1965	1975	1985	Change 1965-1985	Percentage Change 1965-1985	Growth Share 1965-1985
Large urban	22,589	26,682	34,034	11,445	50.67%	34.68%
Suburban	5,628	7,986	11,863	6,235	110.77%	18.90%
Small urban	8,207	10,863	14,611	6,405	78.04%	19.40%
Exurban	7,631	10,311	14,130	6,498	85.15%	19.70%
Rural	3,152	4,399	5,567	2,415	76.63%	7.30%
Total	47,207	60,242	80,206	32,998	69.90%	100.00%

SOURCE: Data from *County Business Patterns*, U.S. Department of Commerce (1965, 1975, 1985).

TABLE 2.3. Distribution of Manufacturing Employment by County Class, 48 Contiguous States: 1965-1985 (figures in thousands)

County Class	1965	1975	1985	Change 1965-1985	Percentage Change 1965-1985	Growth Share 1965-1985
Large urban	7,590	7,126	7,127	(463)	−6.09%	−26.19%
Suburban	2,261	2,359	2,625	364	16.12%	20.63%
Small urban	3,184	3,453	3,503	319	10.02%	18.05%
Exurban	3,437	4,272	4,507	1,070	31.13%	60.56%
Rural	1,039	1,328	1,515	476	45.79%	26.94%
Total	17,511	18,539	19,277	1,767	10.09%	100.00%

SOURCE: Data from *County Business Patterns*, U.S. Department of Commerce (1965, 1975, 1985).

TABLE 2.4. Population Distribution in 48 Contiguous States by County Class, 1960-1985 (figures in thousands)

County Class	1960	1970	1980	1985	Change 1960-1985	Percentage Change 1965-1985	Growth Share 1965-1985
Large urban	63,648	71,590	74,719	77,301	13,654	21.45%	24.46%
Suburban	20,810	26,425	29,391	30,899	10,089	48.48%	18.08%
Small urban	30,312	34,850	39,479	42,182	11,870	39.16%	21.27%
Exurban	42,471	46,882	56,277	58,980	16,509	38.87%	29.58%
Rural	20,277	20,431	23,472	23,965	3,688	18.19%	6.61%
Total	177,518	200,179	223,337	233,328	55,810	31.44%	100.00%

SOURCE: U.S. Census of Population and *Current Population Reports* (1960, 1970, 1980, 1985).

We found that the role of labor influences is impressive. No less than four of the seven significant factors (and five variables, including two for education) involve labor considerations. The factors are (a) existing manufacturing, (b) unions, (c) education, and (d) farm population. Firms avoid existing manufacturing employment concentrations because of overly competitive labor markets. They avoid union labor because of work rules, high wages, strikes, and so on. They seek educated workers because of the workers' productivity. They appreciate labor from farm households because of a perceived better work ethic, possession of mechanical skills, and the hostility of farmers to unionization.

We observe that the competition for the relatively few new plants that may locate in exurban areas at any given time will result in winners and losers. How can exurban communities improve their attractiveness to manufacturing firms? Many determinants of manufacturing growth cannot be changed, at least not very much. There are, however, two or three determinants that exurban places can influence. One determinant is education. High local public education expenditures and percentage of adult population with a high school education is associated with high manufacturing growth. Another determinant is property taxes. Reducing or waiving property taxes may help attract industry. Low property taxes per capita are associated with high manufacturing growth. We suspect that, even in exurban places with high property taxes per capita (indicative of places with high local investment in public education), reducing or eliminating property taxes on manufacturing may be effective in attracting new plants. For some exurban places, a third determinant that can be influenced may be airports. A new airport can help, because less distance to a small airport is associated with high manufacturing growth.

The following example is illustrative. During the 1980s the Midwest attracted six major Japanese auto manufacturing plants and 200 transplant parts suppliers that employ more than 50,000 workers and produce more than 2 million units annually. Where did they locate and why? Generally speaking, they located in the exurban areas of Kentucky, Illinois, Indiana, Michigan, Ohio, and Tennessee[1] (Florida et al., 1988). They also located where we expected: places of low metropolitan manufacture value added, low manufacturing employment as a percentage of total employment, low to moderate union membership, low property taxes (occasioned by tax concessions), and presence of a small airport, relatively high adult population percentage of high school graduates and expenditures in public education, and high farm population.

How can these factors be combined to the advantage of particular communities where the hinterland meets the urban field? Consider the case of speculative industrial buildings in two exurban Georgia counties.

Case study. In recent years, local governments have facilitated the construction of speculative industrial buildings, especially in underdeveloped regions (Falk et al., 1980). A speculative building is erected without a tenant or a buyer in the hope that it will be purchased or rented by a firm in the future. Speculative industrial buildings offer firms the ability to buy or lease space that is not only in place but immediately available. Firms are relieved of the time-consuming work of real estate acquisitions, potential zoning problems, and dealing with architects, lawyers, and engineers. Firms are assured of all basic utility services (Fernstrom, 1979). The popularity of speculative industrial buildings has increased throughout the years with a strong presence in the Southeast and the New England states. Some surveys indicate that up to two-thirds of all industrial location or relocation prospects are looking for immediately available space such as that provided by speculative industrial buildings (Southern Industrial Development Council, 1985). From a public policy view, speculative industrial buildings are meant to induce economic development that may not occur otherwise. Forms of public participation include special financing, extension of facilities and services, land write-downs, and joint ventures with private developers. The public risks losing resources if buildings are not occupied. Speculative industrial buildings are usually local endeavors. They require substantial financial commitment and a willingness to accept risk. In one sense, they represent a community's underlying commitment to a particular kind of economic development. A review of the speculative industrial building policies of Hall and Troup counties, Georgia, is instructive.

Hall County is situated within the Atlanta, Georgia, metropolitan statistical area. Its 1980 population was about 77,000. Its principal city, Gainesville, is about 55 miles from downtown Atlanta. Its 1980 population was about 15,000. By the late 1970s Hall County and Gainesville city officials were concerned that growth in manufacturing employment was lagging behind other parts of the region. They also knew that some firms favorably inclined toward locating in the area could find no immediately available space and located elsewhere. Hall County officials wished to make the county *more* integrated with the Atlanta metropolitan area. It thus embarked on a program to lure firms into the county.

Hall County's speculative industrial building program began with a donation by the Johnson & Johnson Company of 200 acres for industrial development. The county spent $2.8 million to improve the industrial park. The money came from sale of industrial development bonds secured by a one-mill property tax surcharge if land sales do not retire the debt. Since 1981 it has accommodated nearly 1,000 new manufacturing jobs. Tenants include food processors, a beverage distributor, an electronics assembler, a printing company, and an automotive parts distributor. Hall County's program is effective. It has yet to use the dedicated millage to pay IDB debt service.

In contrast is Troup County, Georgia. It is adjacent to the Atlanta MSA and the smaller Columbus, Georgia, MSA. Its 1980 population was about 53,000. Its principal city, LaGrange, is the same distance away from downtown Atlanta as Gainesville in Hall County, or about 55 miles. Its 1980 population was about 26,000. Both Hall and Troup counties had roughly similar industrial development performance between 1975 and 1980. Yet, Troup County officials chose not to participate in the speculative industrial building program for reasons principally relating to avoiding financial risk. This attitude reflected the antitax, antigovernment sentiments of the population of the time. Between 1980 and 1990 Hall County outperformed Troup across all economic development dimensions such as per capita income and manufacture value added.

Troup County, however, decided to implement a speculative industrial building program in the late 1980s. Troup County officials saw that economic development was occurring in other counties that had vacant industrial space. After the decision was made to embark on the program, the Troup County Local Development Authority created an industrial park in LaGrange. LaGrange donated 135 acres of land to the LDC, which the LDC used as collateral to obtain another 500 acres. It later added 400 acres to its inventory. Improvements on the first 135 acres and shell buildings cost about $800,000. Capital came from a local consortium of banks. Land, grading, utility extension, and construction of shell buildings on the 500-acre tract cost $1.2 million. Two million dollars in funds came from IDBs secured by a two-mill property tax. The tax has yet to be assessed. The program has attracted distribution, assembly, and packaging firms that employ nearly 1,000 workers.

The Development-From-Below School

Markusen (1987) observes that the history of U.S. domestic development policy follows two dominant but parallel paths: (a) territorial expansion and consolidation in the process of building a nation regardless of regions and (b) building regions based on achieving homogeneity of political-economic and social orientation within them. As seen above, nation-building policies have been effective. Region-building policies, however, may be misdirected in their aim to create homogeneity among regions. The objective of the development-from-below school is to tailor development patterns to fit regional character.

There is another angle to the development-from-below school. Because they are applied at a macro level across all regions, some development-from-above policies will fail to be effective in certain regions but will be effective in others. For example, steel import quota policies that support steel industries in the Midwest result in higher costs imposed on aerospace industries in the West. Oddly, it is the International Trade Institute of Japan that recommends U.S. development policies be tailored to its distinct regions (Hansen, 1988).

According to Stohr and Taylor (1981), development-from-below involves controlling Myrdal's backwash effects of development-from-above. Development-from-below aims for *generative* growth (Richardson, 1973). Several efforts can be made to reduce importation of goods and services beginning with the most basic of imports. Regionally created savings can be reinvested within the region. Small-scale labor-intensive industries can be fostered. Much of this effort can be coordinated through decentralized substate regional administrative organizations supported by local and state governments and by business groups. The development-from-below school includes three approaches: territorial development, functional development, and agropolitan development. There are some critical flaws in the approach. If properly addressed, however, this school shows promise for attaining the social, political, and economic outcomes desired for particular regions (Clavel, 1983).

Territorial Development

Territorial development theory views the role of regional planning as accommodating if not accelerating the spread of development from growth centers to the hinterlands (Weaver, 1984). Regional development before World War II was viewed as dependent upon growth centers. Regional planning was viewed in part as improving interaction between growth centers and their hinterlands. Regional development is attained through a system of growth centers that attract investment and produce exports, and in turn further decentralize new investment to lagging areas in a kind of accelerated filtering-down process (Berry, 1972). Territorial development is achieved through the integration of growth centers with their hinterlands (Stohr, 1981).

Functional Development

Functional development theory applies a proactive approach to regional development (Friedmann and Weaver, 1979; Stohr, 1981; Markusen, 1987). Its principal assumption is that regional development can be achieved by the harnessing of selected regional resources to create the kind of generative growth envisioned by Richardson. The Tennessee Valley Authority of the Southeast and the Bonneville Power Administration of the Northwest are two examples of functional development. In both cases, major waterways were harnessed both to create cheap electrical power and to make river transportation more efficient. Partly as a result of these efforts, considerable capital was invested in these regions, industries built, migration stimulated, and overall economic well-being enhanced. On the heels of TVA and BPA successes came other programs offered by the core to stimulate regional development in the periphery areas of Appalachia, the Ozarks, the Four Corners, and other economically distressed regions (Markusen, 1987).

Territorial development is concerned with moving periphery or hinterland regions from lower to higher stages of development through a filtering-out process. Functional development is chiefly concerned with moving a region to higher stages of development by organizing it around a principal function.

Agropolitan Development

Territorial and functional approaches to development do not consider the extent to which regions themselves may take control over their social and economic conditions. A major criticism of both approaches is that they require domination of lagging regions by leading regions. Territorial development is dependent on the vitality of growth centers. The assumption is that if a center falters its hinterland fails.[2] Functional development is effected by decisions several thousand miles away (Congress) and, while a region may be improved, it remains dependent upon external regulatory and economic constraints.

Agropolitan development theory (Friedmann and Weaver, 1979) allows regions themselves to determine the kind of development that is desired in economic and social respects.[3] Agropolitan development has its roots in utopian thought, particularly from the writings of Odum and Moore (1938) on cultural regionalism. They argue that only by preserving cultural regionalism would a regional society have a chance to survive industrialization with its tendency to force cultural conformity with the industrial interests. They propose regional-national social planning that would formally create a patchwork of autonomous *organic territories* that would be defined by natural resources, climate, certain historical elements, cultural traditions, and social structure.

Until the middle 1970s, this particular regional development approach failed to influence regional development policy through either the territorial development or the functional development stages. By the middle 1970s, however, there emerged some conscious efforts to consider regional development as fostering not just leading-lagging interaction but fostering regional social and political identities reinforced through a reordering of economic linkages. This is the principal focus of the development-from-below school of regional development.

The agropolitan approach was originally intended to be applied to the hinterlands of Third World countries by Friedmann (1979) and Stohr (1981). Its principal features are applied to the United States by Friedmann and Weaver (1979), Weaver (1984), and Clavel (1983). The proposed organization of an agropolitan regime is based on common cultural, political, and economic spaces that could be reasonably defined spatially. To assure reasonable face-to-face decision-making opportunities, the agropolitan regime is organized to create agropolitan districts ranging in size from about 20,000 to about 100,000; cities are organized similarly into agropolitan neighborhoods.

There is insufficient space here to present all the facets of the agropolitan development concept but some of its principal features can be reviewed. One feature is the assuming of greater control over natural resources. Renewable resources are managed to sustain yields and thereby assure a permanent economic base. In less developed areas with surplus labor and insufficient capital, labor-intensive technologies are used to exploit resources. Efforts would be undertaken to provide goods and services locally that are otherwise imported. This may involve targeting training, incubation, and subsidization policies to specific industries. Highway systems would be built to improve accessibility within agropolitan areas and not necessarily between agropolitan areas and growth centers. Only those export-based industries that raise local quality of life are encouraged. Export-based industries that exploit local resources but do not measurably improve quality of life would be discouraged if not prevented. An effort would be made to create a social consciousness whereby sacrifices by individuals are not seen necessarily as altruistic but as producing a common benefit that is shared by all who sacrifice. One example may be investing local savings into local opportunities even where investment returns may be less than returns earned by investing outside the region. There would also be creation of a process for social learning designed at once to reinforce local cultural and social orientations and to prepare individuals for continuing challenges.

There are serious limitations to the agropolitan approach. While certain aspects of the approach may be more-or-less feasible, many are not. Take, for example, the proposition that an agropolitan district may assume greater control over its natural resources, especially to manage renewable resources to sustain yields and thereby assure a permanent economic base. In many parts of the country, the natural resource base is owned or in various ways controlled by government agencies beholden to interests in other regions. For example, preservation of the Spotted Owl in the northwestern forests satisfies the interests of the preservation-minded in every region but where the loggers and the owls live.

The agropolitan approach would result in highway systems designed and built principally to improve accessibility within agropolitan areas and not necessarily between them and growth centers. This is an expensive and, alas, fruitless proposition. Improving highways within the hinterlands does little to improve regional development prospects. Only highways that improve accessibility of hinterland areas within about four hours' drive from major urban markets are found to be economically beneficial (Forkenbrock, 1990).

The agropolitan approach would require reinvestment of local savings into local enterprises even if returns are less than offered elsewhere. Perhaps in situations where capital is not mobile, this may be accomplished but it is difficult to imagine such behavior on a mass scale in situations where there are few barriers to capital movement. Moreover, as Keynesian economists would point out, mass investment in opportunities generating fewer returns than other opportunities will ultimately leave society worse off; indeed, the purpose of making capital increasingly mobile is ultimately to upgrade the welfare of world society.[4]

Finally, the agropolitan approach presumes a more-or-less homogeneous population. This implies the agropolitan districts would be free to decide who can or cannot live in the district. This, of course, violates the very principles of the U.S. Constitution.

The Evidence and Assessment

Development-from-below policies aimed at developing regions of the United States are guided by both economic and political considerations. For example, the Economic Development Administration was formed in 1965 to funnel federal revenues principally for public works projects in distressed rural regions. But, by the end of the 1990s, the EDA's programs reached jurisdictions representing about 80% of the nation's population including such "distressed" areas as some of Atlanta's burgeoning suburbs. The problem of course is that what is considered distressed in some parts of the country may be considered advantaged in other parts. Hansen (1988) argues for redirecting federal regional development policies to be sensitive to regional heterogeneity.

But the point of the development-from-below school is that regions themselves must have substantial say in determining the character of regional development. This is not without limitations. Development-from-above forces are powerful and they can be viewed as exogenous factors. Those factors can undermine development-from-below efforts or they can be used to fulfill development-from-below efforts. Consider attempts by two agropolitan regions (using the term loosely) to influence development their way.

Washington County, Georgia. This is a rural county clearly in the hinter-lands of Georgia. It is located about 140 miles from Atlanta. Its 1990 population was about 20,000. Its principal city, Sandersville, had a 1990 population of about 7,000. There was virtually no change in population from the previous decade. The economic base of the county is kaolin, which is projected to be exhausted within one to two generations.

The county wants to diversify its economic base. One attempt to do so was the construction of a speculative industrial building. Seeing the success of such buildings around exurban Atlanta, Washington County and Sandersville presumed that "if you build it, they will come." The building was built but *they* did not come. Putting a happy face on an expensive undertaking, local officials see the vacant speculative industrial building as a visible demonstration of their commitment to economic development.

Yet, just as the speculative industrial building is vacant, proposals to reduce "leakage" in the county's restaurant business have been denied. Washington County is "dry" and so all liquor-by-the-drink trade occurs in adjacent "wet" counties. The county loses about $5 million annually in restaurant sales to other counties. It also results in county residents being injured or killed when they return from those counties, driving along narrow, winding roads. An obvious way to stimulate some development is to change county legislation to allow liquor-by-the-drink. This proposal, however, is routinely defeated at the polls. This is one example of agropolitan expression at work. The county chooses less economic activity and greater injury to its citizens rather than sacrificing cultural and social principles.

Oregon's Willamette Valley. Oregon's Willamette Valley contains 10% of the state's land area, produces 40% of the state's agricultural products, and houses 80% of the state's population and employment. The competition for land favors urban development because urban externalities reduce the productivity and hence the value of farmland and because public subsidies for urban development greatly exceed subsidies for agriculture (Knaap and Nelson, 1992). During the 1960s and 1970s, the Willamette Valley was one of the nation's fastest growing regions. In 1973 alone, more than 30,000 acres were lost to urban development. While the state's gross domestic product was increasing, its agricultural productivity was declining. Largely in response to public concern over loosing the open space and economic benefits of preserving farmland, the Oregon legislature passed the nation's most sweeping land use planning laws. These laws require that all urban development be diversified to areas inside urban growth bound-aries and that prime farmland be used only for agricultural activities. The contribution of urban growth to the state's gross domestic product remains but overall domestic product is enhanced because the state has found a way to preserve prime farmland for productive agricultural uses (Nelson, 1992). Since a debilitating recession in the early 1980s, Oregon has become one of the nation's most stable economies in large part because its economy is clearly diversified in manufacturing, shipping, services (including tourism), and agriculture. More important, residents of the valley and the state as a whole expressed their preference to maintain farms (and nearby forests) even if it means higher urban densities.

These examples suggest that development-from-below approaches may have an important role in influencing the nature of development-from-above forces and in guiding development patterns to preserve particular cultural and social principles.

Summary of the Two Schools

Development-from-below fundamentally aims to create regional autonomy through integration of all aspects of life within a territory defined by its culture, resources, landscape, and climate. In contrast, development-from-above aims to achieve functional integration wherein leading regions expand into lagging regions and resources of lagging regions are made more accessible to leading regions (Weaver, 1981). There may be much overlap between the schools. Both schools apply economic base approaches to describing regional development and use economic base concepts to design regional development policies. Both schools acknowledge stages of growth in regional development. Where they diverge is in the level of integration desired between leading and lagging regions and in the capacity of any given region to tailor policies and resources to achieve indigenously determined social, economic, and political institutions.

Implications for
Regional Development Professionals

On the one hand, there has been a significant redistribution of development since World War II away from core regions of the Northeast, Middle Atlantic, and Midwest to the periphery regions of the South, Mountains, Southwest, and West. On the other hand, there has been a parallel redistribution of development away from growth centers in all regions to the hinterlands—or, rather, to the exurbs. This has important implications for regional development professionals.

Functional integration of leading and lagging regions has been achieved through a combination of policies and economic trends. Does this undermine the efficacy of territorial integration of the kind advanced by agropolitan and similar concepts? One may suppose so. The rise of the Sun Belt and of the West has meant greater integration of the periphery with the core; it has also meant the integration of the hinterlands with centers. Does the success of functional integration undermine agropolitan development? Not entirely: What may be under way is a process of blended functional integration with regional consciousness building.

Significant advances in commuting, teleworking, and telecommunication will make the forces of development-from-above more influential in the development of the periphery and especially of hinterlands. Millions more households will work in edge cities or suburban activity centers—which owe their emergence to deconcentration policies of the federal and state governments—and commute 20 to 50 miles to where they live in a kind of splendid isolation (Herbers, 1986; Nelson, 1992). Teleworking will enable millions more households to stay at or near home to engage in work no matter where they live. Improvements in microwave, fiber-optic, and satellite telecommunication technologies will allow millions of "back office" jobs to locate in nontraditional locations ranging from neighborhood-based multifirm-managed employment centers to remote hinterlands.

In this kind of world, hinterlands become reasonable locational alternatives for millions of households. As hinterlands become more settled, certain economic activities follow. Smaller towns benefit from greater income brought into the hinterland. For example, Mitchelson and Fisher (1987) find that long-distance commuting is a form of export economic base for those communities that "export" their attractiveness and "import" households who earn urban incomes. In the future, there may be competition among hinterland areas producing and "exporting" amenities to attract such households and their urban incomes.

Blumenfeld (1983), Lessinger (1986), and Nelson and Dueker (1990) observe that virtually all exurban and hinterland areas of the nation have opportunities for development because of commuting, teleworking, and telecommunication. Sometime in the next century, the contiguous 48 states may be seen more as a mosaic of low-density settlements punctuated not by growth centers from which development emanates but as social, cultural, and governmental centers satisfying certain needs of the hinterland population.

In this world, periphery regions and hinterlands will be playing on a more level field. They may be able to create a unique sense of physical, social, and economic place that is attractive to people of similar orientation. The Hood River Valley in Oregon owes its economic health to a combination of long-distance Portland commuters, a vibrant fruit industry, specialized sporting opportunities involving the Columbia River and Mt. Hood, and a large number of affluent households whose work is done almost entirely in their homes. Similar kinds of people are attracted to the unique social, cultural, and environmental opportunities offered by the rural counties of California, the Georgia and North Carolina mountains, Colorado's mountains, the Kentucky Bluegrass, the Flint Hills of Kansas, upstate New York, and the small towns of New England. Technological advances combined with the ability of millions of people to live nearly anywhere to earn a decent income perhaps will make it easier for an agropolitan approach to regional development to emerge. Indeed, hinterlands that become organized along agropolitan approaches could find themselves more attractive places for people in search of life-styles that offer social and cultural stability in a pleasing environment. The challenge to regional development officials may be more in creating greater regional identity than in chasing after every economic development opportunity no matter what its contribution to regional quality of life.

Notes

1. Also located there is a Volvo plant. By the mid-1990s, BMW will be operating a plant in exurban South Carolina in a location halfway between Atlanta and Charlotte.

2. The evidence of center-to-hinterland population and manufacturing shifts stimulated by development-from-above policies seem to undermine this assumption, however.

3. The reader may equate this term with the economists' concept of "antarchy," the Quebec secessionist movement, and the black nationalists' argument for the control of a sovereign black nation.

4. The California Public Employees Pension Fund is one of the five largest concentrations of capital in the world. Its trustees demand that at least 10% of its investments be made in California. This is a kind of agropolitan investment policy. In recent years, California has seen a substantial decline in growth coupled with reductions in state and local spending despite higher unemployment and rising health care needs. It is possible that the pension portfolio has lost substantial value in part because the fund's very investments helped lead to overbuilding and ultimately the collapse of the real estate market in certain areas of the state. It may continue to invest to "prime the pump" of economic development but its portfolio may continue to lose value. The real beneficiaries of pension fund investment policy during such times are entrepreneurs—many of whom are headquartered in other regions—and not necessarily pensioners.

References

Balassa, B. (1961). *The Theory of Economic Integration.* Homewood, IL: Irwin.

Berry, Brian J. L. (1972). "Hierarchial Diffusion" in N. M. Hanson (ed.), *Growth Centers in Regional Development* (pp. 108-138). New York: Free Press.

Berry, Brian J. L. and Quentin Gillard (1977). *The Changing Shape of Metropolitan Areas.* Cambridge, MA: Ballinger.

Blackley, Paul R. (1986). "Urban-Rural Variations in the Structure of Manufacturing Production." *Urban Studies* 23: 471-483.

Blumenfeld, Hans (1983). "Metropolis Extended." *Journal of the American Planning Association* 52 (3): 346-348.

Bourne, L. S. (1980). "Alternative Perspectives on Urban Decline and Population Deconcentration." *Urban Geography* 1: 39-52.

Clavel, Pierre (1983). *Opposition Planning*. Philadelphia: Temple University Press.

Cohen, Yehoshua S. and Brian J. L. Berry (1975). *Spatial Components of Manufacturing Change* (Research Paper 172). Chicago: Department of Geography, University of Chicago.

Falk, Lawrence H., Darl A. Hellman, and Gregory H. Wassall (1980). *State Financial Incentives to Industry*. Lexington, MA: D. C. Heath.

Fernstrom, John R. (1979). *Bringing in the Sheaves: Effective Community Industrial Development Programs*. Corvallis: Oregon State University Extension Service.

Florida, Richard, Martin Kenney, and Andrew Mair (1988). "The Transplant Phenomenon." *Economic Development Commentary* 12: 3-9.

Forkenbrock, David (1990). *Road Investment to Foster Local Economic Development*. Iowa City, IA: Midwest Transportation Center.

Friedmann, John R. (1966). *Regional Development Policy: A Case Study of Venezuela*. Cambridge: MIT Press.

Friedmann, John R. (1967). "Regional Planning and Nation Building: An Agenda for International Research." *Economic Development and Cultural Change* 16: 1.

Friedmann, John R. and J. Miller (1965). "The Urban Field." *Journal of the American Institute of Planners* 31 (4): 312-320.

Friedmann, John R. and Clyde Weaver (1979). *Territory and Function*. London: Edward Arnold.

Glasmeier, Amy (1990). *High-Tech Promise for Rural Development*. New Brunswick, NJ: Center for Urban Policy Research.

Hall, Peter (1990). *The Carrier Wave*. New Brunswick, NJ: Center for Urban Policy Research.

Hansen, Niles. 1988. "Economic Development and Regional Heterogenerity: A Reconsideration of Regional Policy for the United States." *Economic Development Quarterly* 2(2): 107-118.

Heckscher, E. F. (1919). "The Effect of Foreign Trade on the Distribution of Income." *Ekonomisk Tidskrift* 21: 497-512.

Herbers, John (1986). *The New Heartland*. New York: Times Books.

Hirschman, Albert O. (1958). *The Strategy for Economic Development*. New Haven, CT: Yale University Press.

Holland, S. (1976). *Capital Versus the Regions*. London: Macmillan.

Knaap, Gerrit J. and Arthur C. Nelson (1992). *The Regulated Landscape*. Cambridge, MA: Lincoln Institute of Land Policy.

Lessinger, Jack (1986). *Regions of Opportunity*. New York: Times Books.

Markusen, Ann (1987). *Regions*. Totowa, NJ: Rowman and Littlefield.

Miernyk, William H. (1977). *The Changing Structure of the Southern Economy*. Raleigh, NC: Southern Growth Policies Board.

Mitchelson, R. L. and James S. Fisher (1987). "Long Distance Commuting and Population Change in Georgia, 1960-1980." *Growth and Change* 18: 44-65.

Myrdal, Gunnar (1957). *Economic Theory and Underdeveloped Regions*. London: Duckworth.

Nelson, Arthur C. (1990). "Regional Patterns of Exurban Industrialization." *Economic Development Quarterly* 4(4): 320-333.

Nelson, Arthur C. (1992) "Preserving Prime Farmland in the Face of Urbanization." *Journal of the American Plannning Association* 58(4): 467-488.

Nelson, Arthur C., William J. Drummond, and David S. Sawicki (1992). *Exurban Industrialization*. Washington, DC: U.S. Department of Commerce, Economic Development Administration.

Nelson, Arthur C. and Kenneth J. Dueker (1990). "The Exurbanization of America and Its Planning Policy Implications." *Journal of Planning Education and Research* 9(2): 91-100.

North, Douglass C. (1956). "A Reply." *Journal of Political Economy* 64: 243-258.

Odum, H. W. and H. E. Moore (1938). *American Regionalism*. New York: Holt.

Ohlin, B. (1933). *Interregional and International Trade*. Cambridge, MA: Harvard University Press.

Perroux, F. (1955). "Note Sur la Notion de 'Pole de Croissance' " [Note on the Concept of "Growth Poles"]. *Economique Appliquee* 7: 307-320.

Richardson, Harry W. (1973). *Regional Growth Theory*. London: Macmillan.

Richardson, Harry W. (1985). "Regional Development Theories" in Harry W. Richardson and Joseph H. Turek (eds.), *Economic Prospects for the Northeast*. Philadelphia: Temple University Press.

Rosenfeld, Stuart A. (1992). *Competitive Manufacturing: New Strategies for Regional Development*. New Brunswick, NJ: Center for Urban Policy Research, Rutgers University.

Rostow, Walter W. (1960). *The Stages of Economic Growth*. Cambridge: Cambridge University Press.

Schumpeter, Joseph (1934). *The Theory of Economic Development*. Cambridge, MA: Harvard University Press.

Schumpeter, Joseph (1939). *Business Cycles*. New York: McGraw-Hill.

Schumpeter, Joseph (1947). *Capitalism, Socialism, and Democracy*. New York: Harper.

Southern Industrial Development Council (1985). *How to Develop a Speculative Industrial Building Program*. Atlanta, GA: Southern Industrial Development Council Press.

Stohr, Walter (1981). "Development From Below" in Walter Stohr and D. R. F. Taylor (eds.), *Development From Above or From Below?* (pp. 39-72). New York: John Wiley.

Stohr, Walter and D. R. F. Taylor (1981). "Introduction" in Walter Stohr and D. R. F. Taylor (eds.), *Development From Above or From Below?* (pp. 1-14). New York: John Wiley.

Tiebout, Charles M. (1956). "Exports and Regional Economic Growth." *Journal of Political Economy* 64: 160-164.

Vernon, Raymond (1966). "International Investment and International Trade in the Product Cycle." *Quarterly Journal of Economics* 46: 113-119.

Weaver, Clyde (1981). "Development Theory and the Regional Question" in Walter Stohr and D. R. F. Taylor (eds.), *Development From Above or From Below?* (pp. 73-106). New York: John Wiley.

Weaver, Clyde (1984). *Regional Development and the Local Community*. New York: John Wiley.

Webber, Michael J. (1984). *Industrial Location*. Beverly Hills, CA: Sage.

Weinstein, Bernard L., Harold T. Gross, and John Rees (1985). *Regional Growth and Decline in the United States*. New York: Praeger.

Wheat, Leonard F. (1976). *Regional Growth and Industrial Location*. Lexington, MA: D. C. Heath.

Wheat, Leonard F. (1986). "The Determinants of 1963-77 Regional Manufacturing Growth." *Journal of Regional Science* 26 (4): 635-659.

Suggested Readings

Blair, John P. and Robert Premus (1987). "Major Factors in Industrial Location." *Economic Development Quarterly* 1 (1): 72-85.

Blumenfeld, Hans (1982). *Have the Secular Trends of Population Distribution Been Reversed?* (Research Paper No. 137). Toronto: Centre of Urban and Community Studies, University of Toronto.

Carlino, Gerald A. and Edwin S. Mills (1987). "The Determinants of County Growth." *Journal of Regional Science* 27 (1): 39-54.

Clavel, Pierre (1983). *Opposition Planning*. Philadelphia: Temple University Press.

Coelen, Stephen P., Robert A. Nakosteen, and Michael A. Zimmer (1988). "An Aggregate Model of Manufacturing Firm Migration." *The Review of Regional Studies* 18(1): 57-61.

Cosson, Philip L. (1988). *Economic Analysis of the Speculative Industrial Building Programs in Hall and Troup County*. Atlanta, GA: City Planning Program, Georgia Institute of Technology.

Erickson, Rodney A. and Michael Wasylenko (1980). "Firm Relocation and Site Selection in Suburban Municipalities." *Journal of Urban Economics* 8: 69-85.

Friedmann, John R. and William Alonso (eds.) (1975). *Regional Policy: Readings in Theory and Applications*. Cambridge: MIT Press.

Glasmeier, Amy (1990). *High-Tech Promise for Rural Development*. New Brunswick, NJ: Center for Urban Policy Research.

Herbers, John (1986). *The New Heartland*. New York: Times Books.

Markusen, Ann (1987). *Regions*. Totowa, NJ: Rowman and Littlefield.

Schmenner, Roger W. (1978). *The Manufacturing Location Decision: Evidence from Cincinnati and New England*. Washington, DC: Economic Development Administration.

Smith, Eldon D., Brady J. Deaton, and David R. Kelch (1978). "Location Determinants of Manufacturing Industry in Rural Areas." *Southern Journal of Agriculture Economics* (July): 23-32.

Stohr, Walter and D. R. F. Taylor (eds.) (1981). *Development From Above or From Below?* New York: John Wiley.

Weaver, Clyde (1984). *Regional Development and the Local Community*. New York: John Wiley.

Weinstein, Bernard L., Harold T. Gross, and John Rees (1985). *Regional Growth and Decline in the United States*. New York: Praeger.

Wheat, Leonard F. (1976). *Regional Growth and Industrial Location*. Lexington, MA: D. C. Heath.

PART II

Space-Based Strategies

3

Applying Theory to Practice in Rural Economies

MARIE HOWLAND

In contrast to cities, rural economies are small, undiversified, and disproportionately comprise a low-skilled labor force. Agglomeration economies, the savings that accrue from the spatial concentration of economic activity, are generally absent. These contrasts to metropolitan economies call for a distinct cast to traditional development theories and imply different theoretical outcomes for rural than urban areas.

Although on an individual level rural economies are relatively uncomplicated, taken as a group they are diverse. Rural towns generally owe their existence to one of the following: agribusiness, family farming, resource extraction, manufacturing, tourism, retirement, government employment, or government transfers; whereas all urban areas owe their economic livelihood to a mixture of manufacturing, advanced services, government, and trade. The source of income of a rural economy shapes its growth trajectory, income level, disparities between rich and poor, quality of jobs, and appropriate development strategies, and different models of growth and development apply to each type of rural community.

Because of the diversity of rural and small town America, five theories including central place, product cycle, cumulative causation, and demand-side and supply-side models may explain the course of rural economies growth and development. These models are conceptualized as a set of transparencies overlaid on a map of the United States. In some rural communities, with a more complex economic history and diverse economic base, all models may overlap. In other instances, only one model may apply.

AUTHOR'S NOTE: This chapter incorporates many insights suggested by Margaret Dewar, Amy Glasmeier, Pauline Howland, Mel Levin, Beverly McLean, Norm Walzer, Wim Wiewel, and Calvin Beale. I am grateful to Bruce Phillips and the Small Business Administration for use of the USEEM data. Of course, I am responsible for any remaining errors.

The intent of this chapter is to draw the links between theory and development practice rather than to present a full description of central place, product cycle, cumulative causation, and demand-/supply-models in all their complexity. Three brief case studies of the development experience of the PENNTAP program in Pennsylvania, Branson, Missouri, and Bonaparte, Iowa, follow and illustrate the application of these five theories to practice. The cases also demonstrate the shortcomings of regional models that leave little room for local self-determination. The Bonaparte, Iowa, case study shows that, even when national and regional structural trends portend decline, local citizen creativity, hard work, and cooperation can chart a different course.

Theory

Central Place Theory

Many towns in rural America evolved as service centers for agriculture-based economies. Central place theory (Christaller, 1966; Losch, 1954; Berry, 1967; Segal, 1977; Evans, 1985) provides a framework for understanding the pattern and trends of places characterized by a population spread evenly over the landscape.

Losch postulated a homogeneous plain, with uniform transportation costs in all directions; an even distribution of inputs; identical production technologies, management efficiencies, and products; and identical consumer tastes and incomes. The sales price of a good increases proportionately with distance from the point of production, as shown in Figure 3.1, because transportation costs rise with distance from the site of production. Seen from above, each firm serves a hexagonal market area in equilibrium because of the packing of circular areas (Figure 3.2). Where production economies of scale are present, costs increase at a decreasing rate with larger market shares. Activities with larger economies of scale serve larger markets and therefore tend to be located in higher order settlements (see Figure 3.3).

Central place theory hypothesizes a hierarchy of settlements. The smallest hamlets include the narrowest range of residential services, most often a grocery, gas station, and cafe (Anding et al., 1990). Frequent trips, the need for face-to-face contacts, and low-level economies of scale characterize services found in communities at the bottom end of the size hierarchy.

According to this framework, all activities found in lower order settlements, plus additional functions, are found in higher order (larger) places. This pattern appears to be supported by empirical evidence, at least for the agricultural Midwest. For example, Anding et al. (1990) found that, in addition to the three activities just mentioned, the next order places, which they call "minimum convenience centers," generally include a restaurant,

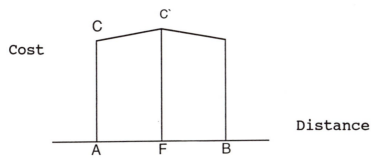

Figure 3.1. Distance From Point of Production

SOURCE: The graphs in Figures 3.1 to 3.3 are adapted from Nourse (1968, p. 53) and Segal (1977. p. 41).
NOTE: Sites A and B are points of production. Line AB is the distance between the two firms. AC is the price at firm A. The delivered price to buyers along the line AB is the price at the firm plus transportation costs. The boundary between the two firms is at F.

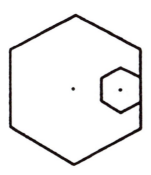

Figure 3.2. Hexagonal Market

NOTE: This figure shows a landscape with two city sizes represented. Each town is the midpoint. The smallest town includes the services with the smallest economies to scale and highest transportation costs. Consumers in the smallest hamlet conduct their frequent purchases locally and travel to the larger town for

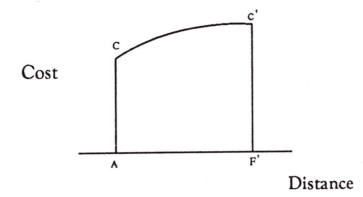

Figure 3.3. Production Economies to Scale

NOTE: With economies to scale, the market reach of firm located at A expands and the firm at B closes. The new market boundary is at F\pri.

bank, hardware store, and drugstore. Accordingly, the full range of service functions are hypothesized to be available in the largest cities. In particular, activities that reach minimum average costs at high levels of output (a football stadium) or require immense market areas to yield sufficient demand (heart transplants) are found only in the largest settlements.

One strength of this model is its power to explain the location of retailing, consumer, and certain types of educational and health services. In spite of steady rural manufacturing growth and a brisk national increase in advanced services employment, rural economies are still heavily dependent on residentiary or nonbasic activities as a source of jobs. Combined, retailing, health, education, and consumer services constitute 40% of total rural employment (Glasmeier and Howland, in preparation).

Although central place theory provides a static view of the rural landscape, with minimal adjustments this model can be made sufficiently dynamic to understand the sources of rural downtown decline, to predict likely winners in the redistribution of residentiary service jobs, and to guide rural downtown revitalization policy. Declining transportation costs, expanding economies of scale, reductions in the intensity and frequency of purchases, and losses in small town populations and incomes all push residentiary services up the urban hierarchy.

Retailing offers a dramatic example. Due to large-scale buying, standardized floor layouts and displays, and computerized inventory processing, retailers are reaching minimum average costs at larger levels of sales than in the past. The result is that retail establishments require larger market areas and as a result are moving up the urban hierarchy. Bigger stores, locating in the larger rural places, offer lower prices and a wider selection of goods. More fuel efficient cars, high-speed roads, and the growing tendency for one member of the farm family to work and therefore shop in larger rural places all reinforce the trend. More dollars spent elsewhere reduce local expenditures, displacing the "mom and pop" enterprises of the smaller towns.

Medical services provide a second example. Hospitals require an increasingly large market area to spread the costs of expensive and specialized equipment. During the 1980s, rural hospitals closed at a high rate or tended to become elderly nursing facilities, as medical services consolidated in the larger nonmetropolitan towns and largest cities (U.S. Congress, 1990). Finally, long-running reductions in transportation costs are extending the distance families are willing to travel for health services (U.S. Congress, 1990).

In the absence of rising incomes and population in the smallest settlements, theory suggests and empirical evidence supports the loss of retailing in the smallest hamlets and slower growth in smaller than larger places.

Data for Kansas, a state likely to approximate this model's assumptions, are shown in Table 3.1.

The central place model dominates other models in a historically important, but increasingly rare, community type—one that derives the major share of its income from labor-intensive farming or spatially distributed mining. In 1975, 29% of rural counties were farming dependent; in 1986, only 21% were farming dependent. In 1975, 6% of all counties were mining dependent, down to 5% in 1986 (Hady and Ross, 1990).[1] Manufacturing is now a larger source of rural income than agriculture, mining, or natural resource extraction. This trend requires that we lay new theories over central place theory in instances where traditional agricultural regions have attracted manufacturing, tourism, or retirees and also discount central place theory in regions whose economic base was never farming or spatially dispersed mining.

Product Cycle

Two critical differences between urban and rural economies define relevant rural development theories. First, recognizing the deficit of external economies is critical for understanding manufacturing-based rural development.[2] Rural manufacturing maximizes profits in the absence of specialized labor, access to informal or recent information about competitors and changing markets, venture capital credit markets, and specialized service inputs. Second, rural economies are generally small and undiversified, implying both a fragility and a potential that are absent in large urban centers. The loss of an industry can devastate a rural economy, while the opening of a single factory can nearly eliminate local unemployment.

Firms that thrive in the absence of agglomeration economies are those that do not require frequent face-to-face contacts with suppliers, markets, and competitors and that do not require a highly skilled or specialized labor force. The product cycle model puts this dependence on unskilled labor into a historical and dynamic framework by hypothesizing that products and processes move through phases. In the initial stage, a product and production process are unfamiliar and changing. Large, complex economies, with their well-developed capital markets, highly diverse and skilled labor force, lower cost access to information, complex constellation of suppliers, and proximity to markets, are most attractive at this phase of the product cycle (Vernon, 1966).

As a product matures, it becomes standardized and its manufacture routine. Establishments can be physically distant from credit markets during this phase of the product cycle because the establishment is usually a branch plant and obtains credit and specialized services through its

TABLE 3.1. Nonmetropolitan Employment Growth, 1980 to 1986, by Town
Size (Annual Average Percentage Growth Rate)

Population	Retail	Consumer	Nonprofit*
25,000 and up	3.25	−.24	12.79
10,000-24,999	2.90	−.34	3.60
5,000-9,999	2.93	−3.09	3.48
2,500-4,999	.13	.18	1.80
2,500 and below	−1.05	−2.70	1.03

SOURCE: University of Maryland analysis of USEEM data, provided by the U.S. Small Business Administration.
*Includes education and health services.

headquarters; from a flexible, skilled labor force, because the work is predictable; from competitors, because the technology and product are not evolving; and from markets, because the product is standardized and frequent client input is no longer essential. These are generally establishments that minimize costs by adopting assembly-line production techniques and using low-cost labor. Rural economies are most likely to capture manufacturing at this later phase of the product cycle.

From the 1950s through the 1970s, rural manufacturing employment growth rates outpaced those of metropolitan areas, a performance that could not be discounted as merely a small increase on a small original employment base (Carlino, 1985). The decentralization of low-skilled assembly-line manufacturing explains this pattern.

The product cycle model provides important insights into the rural renaissance of the 1970s. Events of the 1980s, however, underscore the need for this model's refinement. Today, much of rural America is finding the benefits of low-skilled manufacturing and some export-oriented services to be ephemeral. Improvements in telecommunication and transportation technologies, advancing educational systems abroad, and the willingness of some Third World countries to suppress unions and overlook environmental degradation have facilitated the movement of much of the lowest skilled work offshore. For example, manual data entry, a service industry that behaves much like manufacturing in its locational tendencies, is moving out of rural America to locations in the Caribbean, Philippines, China, as well as other countries. The hourly wages in Barbados are one-fifth those in Appalachia (Howland, 1991). Other manufacturing firms are skipping rural nonmetropolitan America altogether and moving directly offshore.

Second, many products and production processes do not progress neatly through to a maturation and standardization phase. Thus rural America cannot rely on passive development strategies that count on the continuous decentralization of manufacturing. Some products are never mass produced, implying continued urban concentration. Aircraft, for example, continue to be produced in small batches designed to specific client needs and in close conjunction with client specifications. This mature industry has never decentralized.

Third, intense foreign competition from industrialized countries and rising national incomes have shifted some products from mass, assembly-line production toward small batch runs, tailored for smaller, well-defined markets. These flexible production techniques require the more highly skilled, adaptable labor generally found in metropolitan areas and can lead to a recentralization of employment. Barkley and Hinschberger (1992) find this pattern for the metalworking industry.

The adoption of just-in-time techniques reinforces the recentralization of manufacturing. With just-in-time production, firms do not hold large inventories of inputs. The input arrives at the plant just in time to be used in the manufacturing process. Thus product quality must be more predictable than in the past. Both changes strengthen the pull between markets and supplier, implying a tendency for some manufacturing to reconcentrate in traditional manufacturing centers (Schonberger, 1988).

Finally, in other industries, capital is replacing low-skilled labor altogether and negating the need for production to decentralize to regions with abundant, low-cost, nonunion labor. For example, optical scanning is displacing manual data entry—a task often performed in rural America or offshore (Howland, 1991).

The product cycle model draws attention to three sources of rural instability and vulnerability. Rural America has traditionally attracted manufacturing with weak linkages to suppliers and customers. Therefore there is little rural industrial diversification and second-round job creation. Second, rural America has attracted a considerable number of manufacturing firms that located there for the abundance of low-skilled labor. Thus, when the need for low-skilled labor disappears, the firm may have no reason to stay in a nonmetropolitan area. In contrast, firms locate in urban areas for a variety of reasons such as agglomeration economies, diverse labor needs, and proximity to markets. Finally, spatial isolation often leads to a lack of access to new ideas and innovations that might keep rural-based firms competitive in the international arena.

A dependence on manufacturing with weak backward and forward linkages means that rural manufacturing does not act propulsively, as connoted by growth pole theorists (Perroux, 1950). Weak linkages imply little secondary job creation in up- and downstream industries and for the most part have left manufacturing-based rural economies undiversified and subject to cyclical instability. For example, a study of the rural component of the UDAG program found that none of the 100 communities that attracted a manufacturing plant could identify a single additional supplier or customer that had been pulled in as a result (Howland and Miller, 1990). Further recent evidence shows that rural manufacturers have failed to draw advanced producer services, such as banking, insurance, and legal services, in tow. Thus the tradable services, that buoyed the central cities, have not materialized in rural America (Glasmeier and Howland, in preparation). The tendency of rural manufacturing to be in branch plants, that purchase directly from headquarters, further undermines opportunities for second-round job creation in nonmetropolitan areas.

Manufacturing's attraction to rural communities has been for the hardworking, nonunion, and low-cost labor. This has placed rural America in direct job competition with politically stable Third World countries that have literate labor pools. This fact became painfully clear during the recession of the early 1980s, when an overvalued U.S. dollar sent manufacturing jobs offshore. Rural manufacturing employment plummeted. The lack of sectoral diversification makes the cyclical instability of rural manufacturing jobs all the more apparent. Although rural manufacturing growth rebounded during the remainder of the 1980s, rural communities are aware that manufacturing employment is fleeting at worst and unstable at best.

Finally, as a consequence of distance from competitors, markets, and informal urban information networks, rural manufacturers have difficulty remaining abreast of technological change. They often fail to upgrade local skills, adopt new technologies, and find new market niches that will keep them competitive in world markets.

What does product cycle theory tell us about the future of manufacturing in rural America? Clearly, manufacturing is not going to be the panacea for rural revitalization, as it appeared during the 1950s, 1960s, and 1970s. The continuing replacement of routine tasks with machines and the trend toward flexible specialization and just-in-time production will lead to a recentralization of capital to the urban core regions. Continued improvements in telecommunication and travel will facilitate the offshore movement of low-skilled tasks that belie mechanization. All of these trends portend difficulties for rural America.

Neoclassical Theory

The absence of rural economic diversity has important theoretical and practical implications for rural development patterns. In particular, it makes neoclassical growth theory (Borts and Stein, 1964) virtually irrelevant for all but the largest, most diverse rural communities. Neoclassical growth theorists predict declines in investment will be self-correcting as economic deterioration leads to falling wages, rising rates of return to capital, and a rebound in private investment. While the notion of equilibrating markets may apply to urban economies (although the persistence of high unemployment and poverty in central cities challenges this view), the model is irrelevant for small and undiversified economies.

Self-correcting mechanisms are particularly weak in rural economies. The out-migration of labor, the difficulty of assessing risk in a declining region, and quick erosion of existing agglomeration economies lead to further decline, not recovery. In contrast to urban areas, the loss of a major rural employer is devastating and the attraction of a single plant is noticeable in reducing unemployment and increasing incomes. The instability of rural income and employment are conspicuous. Less obvious and measurable is the extent to which investment in supporting services, such as business services, is discouraged by the cyclical variability and fragility of the rural economic base. The circular and cumulative causation model provides a better guide to growth patterns in rural economies.

Circular and Cumulative Causation

The circular and cumulative causation model is compatible with empirical observation. The theory of cumulative causation hypothesizes that, instead of self-correcting, market forces set a direction that is self-perpetuating (Myrdal, 1957: 13-23). Adverse incidents, such as the loss of a major local employer or large-scale farm failures, have repercussions that reinforce economic decline. Similarly, growth stimulating events lead to expansion, which, in turn, continues to attract further growth.

For example, when a plant in a small undiversified economy closes or when another such severe adverse event occurs, workers become unemployed. Aggregate income falls and causes further unemployment in residentiary businesses. Banks tend to overestimate local risk, restrict lending, and cause further economic decay. The lack of alternative job opportunities quickly propels the out-migration of the most skilled workers, making the community a less desirable site for new capital investment. Out-migration also erodes a community's sense of place, reducing local ties and commitment to the community (Bolton, 1992).

In a continuation of the downward cycle, the weakening of a narrow tax base quickly undermines tax revenues. One fallout is a reduction in school and infrastructure spending, further eroding a region's attractiveness to capital. Decline quickly leads to further decline.

In contrast, growth inducing events, such as the opening of a new plant, the discovery of natural resources, or the siting of a government facility initiate a circular and upward cycle. New workers are attracted; population growth stimulates employment in residentiary activities. Increases in population thresholds not only increase demand for existing services but make it feasible for the local economy to support new lines of business. Tax revenues rise, permitting increases in the quality of the local schools and public infrastructure. A better trained labor force may, in turn, attract new investment.

Demand-Side Theory

Demand-side theories, such as Keynesian analysis, export-base theory, and input-output models are more sophisticated than supply-side models, provide insights into one form of rural-based growth, yet are often inadequate for directing rural development policy. Demand-side models hypothesize that regional growth is determined by external demand for the goods produced in a region. The propulsive export activity, in turn, generates income and employment among local suppliers and a residentiary or household sector. Export-base theory is explained in more detail in Chapter 2, and here we focus solely on the model's application to rural economies.

Demand-driven growth models clarify a potential mechanism for rural growth. No doubt, increased external demand for agricultural products, natural resources, and manufactured goods can benefit rural economies. Moreover, the inflow of social security and pension payments act to enhance local growth in the same manner as exported goods. Nonetheless, as guidance for development policy, demand-side theories are limited.

The narrow economic base and branch plant status of most rural manufacturing-dependent economies imply small regional multipliers and weak export-led growth even under the best of circumstances. Rural manufacturers generally perform the low-wage tasks and employ few managers, thus relatively little income circulates through the community. Also, as discussed above, local interindustry linkages are rare and rural residents frequently travel to surrounding towns and cities to shop, resulting in a high propensity to import. Finally, in regions dependent on capital-intensive branch plant manufacturing and agribusiness, some export income may never make it back to the rural community at all.

While multipliers are still likely to be small, the potential for demand driven growth is greater in agricultural than manufacturing or mining counties. Rural economies are more likely to embrace the inputs and markets for agriculture than for manufacturing or mining. For example, seed dealers, farm implement sales, agricultural credit institutions, and food processing plants are typically rural-based activities (Glasmeier and

Howland, in preparation). Still, tendencies for residents to shop out of town undermine the potential of demand driven local development.

Supply-Side Theory

Supply-side models are less articulated but offer greater potential for guiding rural development policy. Supply-side theories attribute increases in local aggregate output to increases in local productive inputs, including capital, labor, entrepreneurial talent, external economies, infrastructure, and natural amenities.

Natural amenities, such as lakes, campgrounds, historical sites, and pleasant climate are critical predictors of which rural communities will perform well in the 1990s. The only nonmetropolitan counties to experience population growth during the 1980s were tourism and retirement counties, and natural amenities are the best predictors of which rural communities have potential as retirement or tourism havens.

In metropolitan regions, local entrepreneurship is an important source of job creation. Spinoffs from existing firms are a major source of new businesses. The lack of dense managerial talent in rural areas inhibits this source of growth. While entrepreneurial possibilities are often deficient in rural communities, areas with natural and climatic amenities continue to attract independent firms. One study of computer services firms found that nearly all of the high-technology computer services firms in rural communities were small independent entrepreneurs, producing for an external market, and had located in rural communities for nonpecuniary benefits (Howland, 1991).

Another less studied supply-side factor is a community's sense of place. This argument is best put forward by Roger Bolton (1992), who argues:

> In some established places there is a sense of community. This sense of community is also capital. It is intangible, and regional economists don't talk much about it, but . . . residents of a place that has a strong sense of place certainly know it and appreciate it. Their appreciation of it is evident by the one bit of evidence that ought to make economists notice: people are willing to pay for it. (p. 192)

Infrastructure investment—in the form of roads, railway stations, industrial parks, sewage and water facilities—lower production costs and have the potential to stimulate rural investment. Yet, decades of unsuccessful experience with these strategies underscore our lack of knowledge about the magic combination of supply-side characteristics that will revitalize rural communities.

Three cases furnish concrete examples of economic theory applied to rural development. The three stories include a state policy to assist rural manufacturing, a flourishing tourism retirement community, and a traditional agriculture community striving to save their downtown. In two instances, theory and market trends suggest economic decline would have resulted in the absence of creative public involvement. The third community was successful with minimal public investment.

Case Studies

PENNTAP: Manufacturing Lead Development

Technical assistance can overcome the disadvantages of distance, lack of information, and isolation for rural manufacturers. PENNTAP is an example. PENNTAP is a Pennsylvania state-funded program designed to provide technology transfer and technological assistance to Pennsylvania businesses and other organizations. Economic development is the program's explicit objective. Started in 1965, the Pennsylvania State University/State Department of Commerce program was designed to disseminate information on new technologies. All services are free; PENNTAP attempts to provide firms with a range of options; and the recipient selects the final course of action. The staff includes engineers, scientists, and librarians. In 1987 nearly 80% (982 out of 1,200) requests for assistance originated from small town and rural Pennsylvania.

The following case exemplifies the accomplishments of a technology transfer program at its best. In 1985 a conductive fiber business in rural northeastern Pennsylvania was having difficulty developing new and marketable products; employment had dropped from 25 to 10 employees. PENNTAP specialists counseled managers on how the business might be able to refine its process for silver-coating synthetic fibers. They also set up some individually tailored computer applications. As a result, the annual business volume increased to $3 million from almost nothing. Because of the success of this project, PENNTAP won first place in a technology transfer competition conducted by the National Association of Management and Technical Assistance Centers (Bartsch and Kessler, 1989: 97).

Manufacturing plants looking for a place to land are rare relative to the number of communities looking for businesses. Therefore supply-side development strategies that enhance local managerial and productive capabilities in existing plants have appeal. This case also demonstrates how public efforts can circumvent the spatial implications of the product cycle model, that is, the tendency for rural areas to depend on low-wage, standardized products and production techniques. In this case, technological change and more sophisticated production techniques were implemented in the rural context. Technical assistance can overcome some of the disadvantages of distance and isolation from centers of innovation and move rural communities away from the futile development strategy that relies on passively waiting for the next industry to decentralize out of urban America. Finally, this case provides an example of demand-side driven growth. The development of a new, competitively priced product led the firm into untapped markets, expanding local incomes and employment through expanded exports.

The previous generation of rural development policy emphasized the attraction of manufacturing. The shortcomings of these efforts are best understood in the context of demand-side and product cycle theories. First, relatively few communities are the recipients of manufacturing enterprises. Second, even when successful sitings occur, a high propensity to import results in small multipliers and insignificant final employment and income growth. Finally, as suggested by the product cycle model, without a more active intervention, rural manufacturing has often provided only temporary and cyclically unstable employment. The lesson is that, where manufacturing chasing strategies bear fruit, supply-side policies, such as technical assistance, worker training, and local infrastructure investments, must occur in tandem to increase the likelihood that rural manufacturing firms will remain competitive in national and international markets.

Branson, Missouri: Tourism and Retirement[3]

Branson, Missouri, is located in the Ozark Mountains. At the turn of the century, Branson attracted a small stream of tourists following up the characters of a widely read book of the period, *Shepherd of the Hills* by Harold Bell Wright. In the 1950s, Branson got a further boost when three dams were finished, creating the Table Rock, Bull Shoals, and Taneycomo lakes. Although built to generate electricity, the lakes attracted sports-oriented tourists interested in fishing, sailing, and water skiing. A local fish hatchery stocks the lakes for sport trout fishing.

A third economic development stimulus was the Marvel Caves, an underground cave system marketed and developed by a Chicago family. As the number of tourists grew, the owners took advantage of idle tourist time and dollars by building a theme park. Constructed on top of the caves, Silver Dollar City now includes a turn-of-the-century village, about 100 crafts people, rides, and live entertainment.

A fourth boost to development occurred in 1983, when county-western singer Roy Clark opened a theater. The combination of the lakes, caves, and Silver Dollar City had drawn a sufficient population to support an audience, and a string of prominent and upcoming musicians performed there. A number of visiting performers, such as Box Car Willy and Ray Stevens, subsequently spun off their own theaters, and Branson has become a center for county-western entertainment. Andy Williams and Shoji Tabuchi opened separate theaters specializing in traditional, classical, and Broadway music. In 1991 Branson attracted 5 million visitors, who spent $1.5 billion (*Time Magazine*, 1992). The 1991 full-year population was 3,706.

Tourism is not the only economic base. The pleasant climate and numerous natural and economic amenities also attract retirees. An estimated 40% of the 1991 permanent population are retirees.

Branson demonstrates supply-side growth, with natural amenities and external economies attracting capital and labor. Lakes, caves, and a pleasant physical environment attracted the original stream of tourists. The potential benefits of agglomeration economies are marked. A concentration of country-western theaters led to external economies, which led to the further spin-off of additional theaters and more performances. The area is now known as a center for new country-western talent, one where new performers are test marketed, and is what *Time Magazine* calls "Country Music's New Mecca" (1992).

Branson also provides an example of the theory of cumulative causation. Once a threshold population size was reached, demand was created for a range of different amenities, such as the Silver Dollar City and country-western theaters. The flow of tourists further attracts a range of services that are in turn attractive to a sizable resident retirement community.

Bonaparte, Iowa: Main Street Revitalization[4]

Bonaparte, Iowa, is a town of 465 people 7 miles from the Missouri border and 23 miles upriver from the confluence of the Des Moines and Mississippi rivers. Like many of the smallest-sized rural towns, Bonaparte has lost population since the turn of the century. The farm crisis of the 1980s was particularly devastating as foreclosures accelerated the loss of farm families, in part replaced by corporate farming. In 1990, 17.3% of the county's population lived in poverty. The loss in population and income, the replacement of the family farm with agribusiness, and the tendency for rural residents to travel further to shop led to the physical deterioration and economic erosion of the downtown.

Central place theory, as well as empirical evidence, suggests that downtown revitalization efforts are difficult if not futile in the agriculturally dependent and declining small town. Increasing economies of scale among retailers and health providers and falling transportation costs undermine the economic viability of the smallest hamlets. Yet the successes of this community demonstrate that coordinated and creative efforts and a strong sense of community allegiance can arrest a downward cycle and rescue a downtown.

One family owned 50% of Bonaparte's downtown retail space and in 1986 announced they were retiring and closing down the local grocery, furniture, hardware, clothing, millinery, and insurance businesses. They insisted their buildings be sold as a block, severely restricting the pool of potential buyers. The block sale in conjunction with the dilapidated condition of the structures meant the death knell for the downtown.

In response, four local residents met with the retiring family and made an offer to purchase the 17,000 square feet of space for $36,000, each putting up $100 in earnest money. To raise the additional funds, the original four devised a plan to sell shares of preferred stock at $2,000 per share. The sales were limited to one share per family to protect families against major loss in the event the venture was not successful and to ensure a broad base of community support. The four lead investors contacted everyone in town and at a communitywide meeting presented the plan for financing the venture, rehabilitating the buildings, and attracting retail stores and services back into the community.

The pitch worked. A nonprofit corporation, Township Stores Inc., was created and within 10 days $100,000 had been raised by selling 50 shares. In 1987, another 16 shares were sold and $250,000 borrowed from the bank.

The plan called for the town corporation to restore all of the buildings, sell some, and lease and manage the remainder. In the original block of buildings purchased by Township Stores, there is now a grocery store, two privately owned condominiums, a school board and supervisors office, a medical clinic with three physicians and a drug dispensary, a hardware/paint store, a beauty shop, gift shop, and two insurance agents. An original private grocery store owner sold out when profits were not sufficient, as did the original owner of the hardware store. Both the grocery and the hardware stores are now managed by Township Stores and are currently operating at a sufficiently profitable level to pay salaries and retire debt. The willingness of town residents to invest $2,000 in an uncertain venture with a below market rate of return is a measure of the economic value of a sense of place and supply-side led growth.

In an example of how accomplishment leads to further success in a small town, there have been a number of beneficial spin-offs from the original revitalization efforts. One outcome was the formation of the Bonaparte Main Street Group, who cooperated to advertise and promote the downtown, raised money to hire an architectural historian to work with volunteers, and successfully added the Bonaparte Historic Riverfront District to the National Register of Historical Places. Bonaparte is currently the smallest Main Street Community in the nation and the first community in Iowa to have the Rural Main Street Designation. Second, Bonaparte received Community Development Block Grants funds for housing rehabilitation. This contributed not only directly to the economy but indirectly as well by creating demand for hardware supplies. Third, the community received a $48,000 state grant to enhance and restore the town's waterfront park, and finally several new stores have opened outside of the Township Stores block. In addition to these directly measurable benefits, the town has received numerous other national and state recognitions. The cumulative causation theory provides a theoretical framework for these achievements.

In the context of central place theory, the commercial successes in Bonaparte's downtown should come at the expense of the surrounding communities. As Bonaparte captures a larger share of fixed, or possibly declining, regional expenditures, other downtowns must inevitably be worse off.

Conclusion and Synthesis

A rich body of microeconomic, macroeconomic, and economic development theory frames our understanding of rural development and underdevelopment. But gaps remain. In particular, the linkages between growth trajectories, economic base, and income inequality are weak. Existing growth theories say little about those in poverty and consistently left behind. The implicit assumptions are that growth benefits all rural residents equally and all individuals are at the mean. In reality, some rural residents don't have the skills to equip them for the labor force even under the best of economic circumstances. The theories contribute little to our understanding of the impact of growth on income distribution or persistent poverty.

Second, theory has been woefully inadequate in predicting the impact of telecommunication on rural communities. Theory long held that rural areas would be the winners in the rapidly evolving information economy. The spread of facsimile machines, teleconferencing calls, video conferencing, and other innovations that reduce the costs of crossing space were to revitalize rural America (Drucker, 1989; Noyelle, 1983). The predicted rapid decentralization of employment, such as back office jobs, has not materialized.

Another arena for theoretical work is in measuring the value of small places to society. The current conservative political climate and federal budgetary crisis have drawn resources away from expenditures construed as redistributive. Yet, there remains a strong, shared national consciousness that small towns, rural America, and place diversity are worth preserving. For example, in spite of the fact that only about 3% of the population farms for their livelihood, the public continues to support massive farm subsidies.[5]

Similarly, subnational supply-side theory needs further development. Macroeconomists are addressing this gap at the national level, but many of the theoretical and empirical breakthroughs are not necessarily transferable to local economies. For example, investments in education are responsible for past national economic growth and necessary for future competitiveness. Yet, investment in nonmetropolitan local education often has the reverse impact, as the best educated leave for the cities and further undermine the region's attractiveness to capital. Aside from anecdotes, we know little about the combinations of natural amenities, capital subsidies, public infrastructure, and labor force investments that can stimulate development in a rural community.

Notes

1. Farming- and mining-dependent counties are defined as having 20% or more of local income derived from that activity.

2. External or agglomeration economies are the savings that accrue when economic activity is spatially concentrated. In urban areas, economies of scale occur due to large markets; a large labor pool reduces search costs; infrastructure is publicly provided; and information can be obtained at relatively low cost. All contribute to lower production costs in urban rather than rural economies.

3. This is based on a telephone interview with Jan Eiserman, Executive Director of the Ozark Marketing Council (June 1992).

4. This is based on an interview with Rebecca Reynolds-Knight, former mayor of Bonaparte.

5. Even more surprising is that public support continues in spite of the fact that large shares of these subsidies go to corporate farming rather than the family farm.

References

Anding, Thomas, John Adams, William Casey, Sandra de Montille, and Miriam Goldfein (1990). *Trade Centers of the Upper Midwest: Changes from 1960 to 1989*. Minneapolis: Center for Urban and Regional Affairs, University of Minnesota.

Barkley, David L. and Sylvain Hinschberger (1992). "Industrial Restructuring: Implications for the Decentralization of Manufacturing to Nonmetropolitan Areas." *Economic Development Quarterly* 6 (1): 64-79.

Bartsch, Charles and Andrew S. Kessler (1989). *Revitalizing Small Town America: State and Federal Initiatives for Economic Development*. Washington, DC: Northeast/Midwest Institute.

Berry, Brian (1967). *Geography of Market Centers and Retail Distribution*. Englewood Cliffs, NJ: Prentice-Hall.

Bolton, Roger (1992). " 'Place Prosperity vs. People Prosperity' Revisited: An Old Issue with a New Angle." *Urban Studies* 29(2): 185-203.

Borts, George H. and Jerome Stein (1964). *Economic Growth in a Free Market*. New York: Columbia University Press.

Carlino, Gerald A. (1985). "Declining City Productivity and the Growth of Rural Regions: A Test of Alternative Hypotheses." *Journal of Urban Economics* 18: 11-27.

Christaller, W. (1966). *The Central Places of Southern Germany* (translation). Englewood Cliffs, NJ: Prentice-Hall.

Drucker, Peter (1989). Editorial. *Wall Street Journal* (April 4): A22.

Evans, Alan W. (1985). *Urban Economics: An Introduction*. New York: Basil Blackwell.

Glasmeier, Amy and Marie Howland (In preparation). *Rural Economies in the Information Age*. College Park, MD: Department of Urban Studies and Planning, University of Maryland.

Hady, Thomas F. and Peggy J. Ross (1990). *An Update: The Diverse Social and Economic Structure of Nonmetropolitan America* (Staff Report, No. AGES9036). Washington, DC: Economic Research Service (May).

Howland, Marie (1991). *Rural Computer Services in an International Economy* (Final Report to the Ford Foundation). College Park: Department of Urban Studies and Planning, University of Maryland (September).

Howland, Marie and Ted Miller (1990). "UDAG Grants to Rural Communities: A Program That Works." *Economic Development Quarterly* 4 (May): 52-63.

Losch, A. (1954). *Interregional and International Trade*. New Haven, CT: Yale University Press.

Myrdal, Gunnar (1957). *Rich Lands and Poor*. New York: Harper & Row.

Noyelle, Theirry (1983). *The Economic Transformation of American Cities*. Totowa, NJ: Rowman and Allanheld.

Perroux, F. (1950). "Economic Space, Theory and Applications." *Quarterly Journal of Economics* 64: 89-104.

Schonberger, Erica (1988). "From Fordism to Flexible Accumulation: Technology, Competitive Strategies, and International Location." *Environment and Planning D* 6: 245-262.

Segal, David (1977). *Urban Economics*. Homewood, IL: Irwin.

Time Magazine (1992). "Country Music's New Mecca." (August 26): 64.

U.S. Congress, Office of Technology Assessment (1990). *Heath Care in Rural America* (OTA-H-434). Washington, DC: Government Printing Office (September).

Vernon, R. (1966). "International Investment and International Trade in the Product Cycle." *Quarterly Journal of Economics* 81: 190-207.

Wright, Harold Bell (1907). *Shepherd of the Hills*. New York: Grosset and Dunlap.

Suggested Readings

Anding, Thomas, John Adams, William Casey, Sandra de Montille, and Miriam Goldfein (1990). *Trade Centers of the Upper Midwest: Changes from 1960 to 1989*. Minneapolis: Center for Urban and Regional Affairs, University of Minnesota.

Barkley, David L. and Sylvain Hinschberger (1992). "Industrial Restructuring: Implications for the Decentralization of Manufacturing to Nonmetropolitan Areas." *Economic Development Quarterly* 6 (1): 64-79.

Bartsch, Charles and Andrew S. Kessler (1989). *Revitalizing Small Town America: State and Federal Initiatives for Economic Development*. Washington, DC: Northeast/Midwest Institute.

Glasmeier, Amy. (1990). *High-Tech Promise for Rural Development*. New Brunswick, NJ: Center for Urban Policy Research.

Hady, Thomas F. and Peggy J. Ross (1990). *An Update: The Diverse Social and Economic Structure of Nonmetropolitan America* (Staff Report, No. AGES9036). Washington, DC: Economic Research Service (May).

Howland, Marie and Ted Miller (1990). "UDAG Grants to Rural Communities: A Program That Works." *Economic Development Quarterly* 4 (May): 52-63.

Walzer, Norman (ed.) (1991). *Rural Community Economic Development*. New York: Praeger.

4

The Economic Development of Neighborhoods and Localities

WIM WIEWEL
MICHAEL TEITZ
ROBERT GILOTH

The practice of neighborhood-based economic development arises out of decades of community organizing and antipoverty efforts. The expansion of urban programs during the 1960s laid the basis for decentralized development activities. By 1970 there were some 200 community development corporations (CDCs) and many other neighborhood groups attempting to improve conditions in small areas. As economic conditions in inner cities worsened during the recessions of the 1970s and 1980s, the number of CDCs expanded to an estimated 2000 at the current time (NCCED, 1991). Housing rehabilitation and construction are the main activities for most of these groups, but they also engage in economic development, which includes assisting business creation and retention, developing commercial and industrial space, and job training programs.

This practice of neighborhood development has been based on such politically charged theoretical concepts as "black capitalism" and "neighborhood empowerment" as well as on mainstream theories drawn from regional and international development. This chapter starts with a review of these theories and their utility in understanding reality and guiding practice. This will lead to a summary of criteria that an adequate theory of neighborhood economic development should satisfy. Because such a theory does not currently exist, the chapter turns to practice next, to discover elements of implied theory in the actual work of neighborhood economic development practitioners. This review of strategies and their relative

AUTHORS' NOTE: The authors acknowledge the advice and comments from Richard Bingham, Marie Howland, and Robert Mier.

success will help in the exploration of an integrated theory of neighborhood economic development.

Current Theories

There are four bodies of theoretical work relevant to understanding neighborhood economic development. The first one is the set of mainstream market-based theories of economic development; the second consists of Marxian critiques and alternatives to these theories; the third derives from political science and focuses on interest groups, political power, and institutions; the fourth is based on sociological theories of community (Wiewel et al., 1989).

The operation of market mechanisms underlies the first set of economic development theories, such as export-base theory, cumulative causation theory, and filtering theories. In all of these, the conditions of individual localities are the outcomes of free choices by firms and individuals maximizing utility based on differing factor prices and their own preferences. For instance, in the analysis by Anthony Downs (1979), the decline of inner-city neighborhoods results from the demand from households and firms for newer facilities and from the desire to live in suburbs. This leaves older buildings obsolete and dilapidated; as demand declines, rents go down, making maintenance impossible. Only when enough growth occurs in the region as a whole will certain neighborhoods be revitalized, depending on their location and physical characteristics.

Within this view, localities are not necessarily seen as entirely helpless. Even though they can't totally change the market conditions they face, they can affect their competitive position by improving business conditions, marketing a neighborhood, or providing incentives of various types. On the whole, though, market-based approaches offer only a limited guide for action and intervention. Moreover, they do not explain differential outcomes for similarly situated neighborhoods. More than just the invisible hand of the market seems to be operating when one neighborhood or small town revitalizes (or declines) and another, similar one doesn't.

In the Marxian understanding of neighborhoods, the main determinant of sociospatial differentiation is the function of residents and places of employment within the overall system of production (Scott, 1980). As David Harvey (1981) argues, residential differentiation reflects the class structure, including not just the division of labor but also consumption classes (e.g., "yuppies"), authority relations, social mobility characteristics, and so on. In this view, areas are poor not because they are lagging but because they are being exploited or are simply superfluous—"lumpengeography," as Walker (1978) called them. They function as a reserve, keeping down the cost of doing business elsewhere and providing a reservoir of workers and land.

Here, too, determination is not absolute. As people and places are sorted, communities form that can become the basis for "displaced class struggle" (Harvey, 1981). This explains phenomena such as the movement that sought to separate black Roxbury from the rest of Boston as the new city of Mandela, as well as the opposition by real estate interests in Santa Barbara against oil exploration off the coast of Santa Barbara, and a host of other neighborhood-based oppositional movements (Kennedy and Tilly, 1987; Molotch, 1979).

An analysis that reduces all neighborhoods and all social movements to reflections of the division of labor and class struggle is too deterministic, however, and gives insufficient weight to the independent effect of politics, culture, and social structure (Castells, 1983). Furthermore, if the class structure is as fine-grained as discussed by Harvey (1981), then the traditional categories of "capital" and "labor" wind up including so many competing groups as to make the result look like regular political pluralism. Thus the political process is clearly of great importance in understanding the differential fates of neighborhoods.

In the third set of theories, the dominant conceptualization of the relationship between the political process and the process of urban growth and neighborhood differentiation is the concept of the "growth machine." Especially during the 1950s and 1960s, this coalition of local government, unions, and place-bound business interests (newspapers, real estate developers, utilities, department stores, banks) pursued its agenda of downtown renewal, economic shift to the service sector, and creation of safe (upper-) middle-income enclaves in increasingly minority-dominated and poor central cities. But, given the central role of local government in this agenda, and given the consequences of urban renewal, highway construction, and suburbanization of employment for poor and working-class neighborhoods, the local political arena became increasingly contested as racial and ethnic minority populations grew and whites moved out (Mollenkopf, 1983; Clavel and Wiewel, 1991). Previously excluded groups sought participation through a variety of protest movements and the development of neighborhood organizations during the 1960s and, especially, the 1970s and 1980s. This led to some level of bureaucratic enfranchisement, in the form of participation in advisory bodies and such new institutions as community planning boards (Fainstein et al., 1983). In some cities it went even further, as progressive mayors were elected (Clavel, 1986; Clavel and Wiewel, 1991). Even in these cases, though, local governments turn out to be quite limited in their ability to change drastically the course of neighborhood development (Wiewel and Rieser, 1989). While the political arena is obviously a central focus for the struggle over development directions, it is still only one part of the picture.

The final body of literature relevant to an examination of the theoretical basis of neighborhood economic development rests on the long sociological tradition of community analysis. Its development over time was nicely summarized by Wellman (1976). He argues that during the 1950s and 1960s the idea of community as a cohesive geographic unit lost its preeminence to theories of mass society. Ethnographic studies, however, subsequently led to a renewal of perhaps romantic pictures of neighborhoods. He then goes on to reconceptualize "community" as a network of like-minded individuals, which for some people may have an important geographic component but for others transcends spatial boundaries. The relative importance of spatial versus other forms of community obviously differs by life-cycle stage, gender, race, income, and other characteristics.

If neighborhoods exist as forms of community, they can have both positive and negative effects of their own, regardless of how they were first formed. Jencks and Mayer (1990) refer to this as "neighborhood effects," that is, the existence of spatially determined social characteristics. Bolton refers to the positive side of it as the "sense of place," the social capital that makes some places preferable to others:

> Residents in places that have a well-developed sense of community have their own preferences and their own mechanisms—public as well as private—for investing in the creation and maintenance of a sense of place. The social capital is productive and it has local public-good aspects. The critical question for state and national policy is whether the "publicness" extends over a wider range of space than the community itself. (Bolton, 1992: 192-193)

The ghetto pathologies described by William Julius Wilson (1980) and others are well-known examples of the negative form of neighborhood effects. To the extent that these neighborhood effects exist, attempts at neighborhood development need to prevent their negative manifestations and, conversely, make use of them in a positive manner.

The Missing Links

The theories discussed so far are lacking in comprehensiveness. Those that are based on neoclassical economics and deal mostly with the larger economy tend to consider neighborhoods as the accidental result of market processes. On the other hand, those that offer insight into neighborhood dynamics, such as the sociological community theories, tend to ignore the external economic constraints. A good theory of neighborhood economic development needs to include an accurate and complete economic perspective, which explains both internal dynamics and the linkage between neighborhoods and the larger economy. Second, such a theory must be based on a realistic understanding of sociopolitical conditions to gain clarity for action: what the basis for intervention at the neighborhood level can be, and how, when, and where interventions can be effective. Of course, it may well be that the formulation of one single theory would necessarily be of such an abstract character as to be useless; perhaps these criteria need to be applied to a set of interrelated theories.

The next section discusses eight forms of neighborhood economic development practice. After a brief definition, we analyze each by exploring its theoretical basis in regard to the two elements mentioned above: (a) the economic basis of the practice in terms of the neighborhood economy and its linkage to the larger economy, and (b) the assumptions in regard to the normative basis for action and the sociopolitical conditions that make success possible. Table 4.1 summarizes this section.

Neighborhood Economic Development Practice

Business retention involves neighborhood organizations promoting the stabilization of existing businesses and industrial districts (Center for Urban Economic Development, 1987). These businesses are frequently bypassed by market trends and by public actors because of their location, size, sector, or profit level. Neighborhood business retention identifies business problems, organizes business leaders, provides technical assistance and loan packaging, organizes collective services (e.g., security or employment referral), launches industrial real estate projects (e.g., industrial parks or incubators), and advocates for public policies (e.g., industrial protection, land use controls, or specialized loan programs) beneficial to specific industrial locations, economic sectors, and firm sizes (Giloth and Betancur, 1988).

Commercial revitalization involves neighborhood organizations promoting the economic growth of commercial districts by sponsoring marketing campaigns, special service (i.e., taxing) districts, commercial strip management (as at shopping centers), business attraction and retention services, and targeted real estate development. Neighborhood business organizations undertake this work because it represents a common interest that no single business can accomplish.

Business ventures. Neighborhood organizations initiate or facilitate new business ventures because neighborhood locations lack indigenous entrepreneurs and are unattractive to market-driven actors. The assumption underlying this strategy is that local businesses hire locally and spend locally, hence strengthening neighborhood economies. Common ventures include construction companies, cooperatives, property management firms, and recycling businesses (Wiewel and Mier, 1986).

Entrepreneurialism is a variant of the business venture strategy in that it assumes that homegrown entrepreneurs will enhance neighborhood ownership, employment, and development. It differs from that strategy in that it trains and nurtures entrepreneurs, attempting to tap human resources that have been underused (McKnight and Kretzmann, 1990).

Neighborhood capital accumulation involves neighborhood ownership and control over land, businesses, investment, and financial capital because neighborhood resources and opportunities are being drained, disinvested, or neglected (Gunn and Gunn, 1991). Frequently this strategy requires establishing new institutions, such as community development corporations (CDCs), land trusts, community development credit unions, or development loan funds that invest resources in neighborhood development ventures.

Education, training, and placement as a strategy invests in the human capital of neighborhoods and attempts to connect people with jobs. This strategy addresses the mismatch in regard to skills and geography of neighborhood residents and the labor market. Its initiatives provide basic skills, employment training, transportation, job readiness, antidiscrimination efforts, job linkages, and ongoing on-the-job supports.

Labor-based development begins by identifying the employment skills of neighborhood residents, particularly the unemployed and recently displaced workers; it then seeks industries that require similar occupational skills. Industries that meet this criterion and are growing in the region are then analyzed in terms of their location, employment, capital, and business assistance needs. Finally, a plan is developed to address how these industries can be attracted to specific industrial districts and sites and linked with the existing labor force (Ranney and Betancur, 1992).

Community organizing/planning alters the power relationships that constrain the flow of resources and opportunities to neighborhoods. Community organizing/planning mobilizes the power of numbers—residents, businesses, and institutions—to advocate, demand, negotiate, and plan. Frequently, neighborhoods band together in citywide coalitions to maximize their strength and to address systemic biases in the decision-making processes and actual allocations of public and private resources. Typical targets for such organizing/planning are city hall, banks, federal dollars (e.g., CDBG), and big development projects (Giloth, 1988).

(text continued on page 88)

TABLE 4.1. Forms of Neighborhood Economic Development

Neighborhood Economic Development Strategies	Economic Assumptions		Sociopolitical Assumptions	
	Neighborhood	Region	Levers for Intervention	Conditions for Effectiveness
Business retention	Business-neighborhood jobs link Appropriate scale	Neighborhood business part of regional economy	Plant closing	Business characteristics State of economic development organization
Commercial revitalization	Commercial areas linked to neighborhood income	Regional hierarchy of commercial places Competition	Commercial strip decline Effects on residential neighborhood	Modest incomes Slow pace of change
Business ventures	Lack of entrepreneurs Lack of neighborhood $$ Undervaluation of neighborhood	Disinvestment/barriers Potential marketplace	Neighborhood income leaks Plant shutdowns Lack of entrepreneurs	Exceptional markets, people, & organizations
Entrepreneurship	Underuse of neighborhood human resources	Barriers Marketplace	Life skills of residents Lack of alternatives Welfare reform	Ongoing support system Self-reliance
Neighborhood capital accumulation	Neighborhoods have underused/disinvested resources	Disinvestment by larger institutions	Lack of neighborhood organization & financial resources	Level of investment Leadership Scale & nature

Education and training	Underused human resources	Regional economic assets	Consensus on need for educational reform & competitive work force Impact on poor	Private sector role Broader involvement of public/private
Labor-based development	Neighborhood as workplace & residence Economic clustering	Growing regional sectors	Plant shutdowns	Existing industry-neighborhood links Sectoral clustering
Community organizing/planning	Unequal power & resources Strength in numbers and organizations	Targets for change	Inequitable resource distribution	Legal guidelines Strength of coalitions Support of public sector

Linkage to the Neighborhood and Regional Economy

What are the theoretical notions reflected in these neighborhood development strategies? First, what does each strategy assume about the neighborhood economy and its linkage to the larger economy?

Business retention assumes that there is a beneficial relationship (or a potential one) between neighborhood industries, residents, and retail businesses because of proximity. It combines this social benefit orientation with a recognition of the importance of market-based factors of industrial location, such as infrastructure, transportation, access to markets and suppliers, and labor force.

Neighborhoods may also be the most appropriate organizational level at which to gather information, bring together businesses for cooperative purposes, and implement service programs. When organized at the citywide level, such activities are frequently dominated by downtown interests, the largest firms or sectors, or promising high-growth sectors. Consequently, neighborhood industrial councils resemble mini-growth coalitions and comprise landowning firms, banks, utilities, and realtors that promote specific industrial districts.

Neighborhood industries are connected to regional markets, are affected by interindustry and corporate relationships, experience locational and agglomeration advantages or disadvantages, and are the target of competition for economic development between regions. Consequently, if regions lose competitive advantage or experience disinvestment, neighborhood locations within these regions will falter as well. Hence regional economic development policies (and national policies with regional implications) affect the viability of neighborhood business locations.

Commercial revitalization. Neighborhoods are part of a market-based hierarchy of commercial places defined by types of goods, family income, and transportation access. The neighborhood shopping district is home to a mix of convenience and trade goods sold by businesses that depend upon the consumer incomes of neighborhood residents. When consumer preferences are not met in neighborhoods, neighborhood income leaks or drains to other shopping districts. Changes in neighborhood income and demographic characteristics, modes of transportation, consumer preferences, retail efficiencies, and the emergence of shopping malls have disrupted neighborhood shopping districts over the past 20 years. Consequently, neighborhood commercial groups also function like mini-growth coalitions, marketing the viability of specific neighborhood commercial areas.

Neighborhood shopping districts are part of regional hierarchies of commercial areas that are constantly evolving—particularly in response to the decline of city populations and the growth of suburbia. As a result, neighborhood shopping districts compete with other city and suburban retailing centers.

Development of business ventures as a strategy is grounded in the lack of neighborhood entrepreneurs, leaks of neighborhood income, and the undervaluing of neighborhood market opportunities due to racial prejudice.

The insufficient rates of return for neighborhood businesses result from low neighborhood income, high operating costs, and higher thresholds of profit required by larger businesses. This perspective even questions whether a market-defined rate of return is an appropriate measure for business investment in neighborhoods given the theorized social benefits that accrue in neighborhoods from business development but that are not counted on the private balance sheet (e.g., employment effects on crime, social welfare, family life) (CWED, in preparation; Wiewel and Mier, 1986).

Lack of business development and ownership in neighborhoods means that income is drained to the region. Moreover, outside institutions—for example, banks—may undermine the neighborhood business environment by creating a scarcity of capital. Alternatively, business ventures that capitalize on neighborhood locational attributes (e.g., amenities, history, culture) may become regional attractions.

Entrepreneurship shares the basic assumptions of the business venture strategy but is also concerned with underused human resources. That underuse occurs because of diminished opportunities and the lack of sociocultural supports needed to encourage entrepreneurs.

While the larger region may be a competitor for a neighborhood, especially for its commercial businesses, it also serves as a potential marketplace for newly inspired or organized entrepreneurs.

Neighborhood capital accumulation. Even poor neighborhoods have financial, human, and organizational resources that can provide a foundation for neighborhood development (Gunn and Gunn, 1991). Unfortunately, these resources are drained from neighborhoods by absentee ownership, lack of local businesses, and institutional disinvestment. That drain of resources from neighborhoods is often related to race and class.

The neighborhood drain of resources occurs because of gaps in neighborhood economies and because of structured forms of disinvestment (i.e., redlining or deindustrialization) that are directed by regional and national institutions. As a development strategy, neighborhood-based financial institutions have been able to attract social investments from the region.

Employment, training, and placement. Many inner-city neighborhoods are defined, in labor market terms, as redundant, surplus, or marginalized. The roots of this pattern are to be found in the functioning of the labor market in conjunction with the historic legacy of racism. Lack of trained and job-ready workers has prompted business communities in many areas to advocate educational reform. Their impetus is the negative impact of the lack of highly trained, technical workers on the competitive position of the region and their own businesses. Teitz (1989) posited a neighborhood's labor force as its key economic asset and in fact its only developable one. One problem with such a strategy, however, is that it may simply let successfully trained individuals leave the neighborhood.

Labor-based development seeks to reestablish the neighborhood as workplace and residential space. That strategy may work best with a cluster of neighborhoods, or a subsection of the city, because of the workplace/residence separation and the need to have a critical mass of specific skills to attract specific industry. Sectors and firms growing or attracted to the region are the target for the labor-based strategy, in particular, manufacturing sectors that contain smaller firms and are labor intensive.

Community organizing/planning. Low- and moderate-income neighborhoods have unequal political power in relation to major private interests and public sector bureaucracies. Neighborhoods are viewed as secondary in relation to land-based interests that promote downtown and big development projects as the engines for city growth. Community organizing builds upon the neighborhood strength of numbers of people and organizations. It focuses on citywide/regional patterns of power and resource distribution, attempting to increase the recognition of neighborhoods and the responsiveness of regional institutions to neighborhood concerns.

Sociopolitical Conditions and the
Basis for Action and Success

Organizations continue to use these strategies because they rest on plausible theoretical notions about the economy. How well they work in particular situations depends on how well they fit with particular economic conditions. Sound economics is not enough, however. Strategies must resonate with values and they must be appropriate to the social and political situation. What are the assumptions of these strategies regarding the basis for action and the conditions for success?

Business retention. Threats of firms closing or moving may precipitate the involvement of neighborhood organizations. Usually the possibility of losing a neighborhood anchor, or having residents laid off, is a powerful motivator.

Variables related to the scale, diversity, ownership, sectoral clustering, and location of industries on a regional and neighborhood basis affect whether cooperative strategies among neighborhood industries are feasible. The level of economic development organization within the neighborhood and its city also affects whether neighborhood business retention is a viable strategy (Ranney, 1988).

Commercial revitalization. Declining sales, strip deterioration, and changing demographic characteristics have encouraged neighborhood retailers to join together to redefine the focus, management, and marketing strategies for their commercial areas. The negative effects of declining commercial areas on nearby residential neighborhoods has also spurred community action. Successful commercial revitalization requires modest income levels and a controllable pace of neighborhood change. It also works when there are few nearby competitors—suburban or city shopping malls, for example.

Business venture opportunities result from an analysis of neighborhood income leaks, export opportunities, business shutdowns, and sheltered neighborhood markets. Their pursuit is often motivated by the perceived absence of any viable alternatives. They have proven difficult except when there are exceptional markets, human resources, and organizational capacity. Neighborhood organizations frequently do not have the entrepreneurial experience, drive, and resources to make fledgling businesses thrive; their ventures face all the rigors that confront start-up businesses in addition to the challenges of operating in neighborhoods or markets that private entrepreneurs ignore (Wiewel and Giloth, 1988).

Entrepreneurship. Most people have life experiences and skills that are translatable into business opportunities, even when business is defined as a micro business or self-employment. Here, too, the absence of other viable alternatives is an important motivation and so is the American image of the "self-made man." A public sector motivation is the concern for enabling welfare recipients to get off welfare; with few jobs available, and a lack of child care and basic skills, part-time self-employment is a viable alternative.

Most small businesses fail—an outcome that probably also applies to firms initiated by entrepreneurial and micro-business programs as well. These efforts, however, despite the failures, may enable people to become more self-reliant and to learn from failure. In this regard, the most successful programs are those that provide ongoing supports for entrepreneurs; neighborhood organizations have been able to organize such support models (Center for Urban Economic Development, 1987).

Neighborhood capital accumulation. Lack of financial resources for investment in viable neighborhood ventures often precipitates interest in establishing alternative financial or development organizations. That issue is often
highlighted by Community Reinvestment Act (CRA) analyses that show
lack of credit flows to neighborhoods by mainstream financial institutions.

There has been a proliferation of neighborhood-oriented financial institutions; very often, however, these institutions operate in larger environments than just single neighborhoods. At the neighborhood level, capital
accumulation has been aided by community development corporations,
through their development and ownership of housing and other real estate.

Employment, training, and placement. The employment needs of the business
community and the deplorable state of public education in U.S. cities have
created the policy space for the overhauling of educational bureaucracies,
experimentation with popular education at the school, family, and neighborhood levels, and innovative job training, placement, and support programs. The differential impact on minority and poor communities has also
provided a touchstone for community outcry.

These initiatives are complicated; their success for any neighborhood
depends upon the strength of the contacts with the private sector, the
stability of the job base, and the support for basic skills, job readiness, as
well as ongoing support. In a broader sense, overhauling the education and
welfare systems that inhibit quality employment and lifelong learning is a
long-term strategy that reaches beyond individual neighborhoods.

Labor-based development. A catalyst for this strategy is a plant shutdown
or a series of shutdowns that affect sectoral clusterings of firms (e.g., steel,
fabrication). The aftermath of these events is a massing of displaced
workers with accumulated skills. Retraining is one option to get them back
in the labor force; the labor-based approach designs development strategies based upon their existing job skills.

This strategy attracts higher wage jobs to neighborhoods by offering an
array of inducements, including labor force training and business support
services. It is most effective where there are industrial districts that already
employ neighborhood people and have locational and infrastructure attributes attractive to a broad range of firms.

Community organizing/planning. The ongoing motivation for community
organizing is the inequitable distribution of resources to neighborhoods
and the social and economic impacts related to that distribution. Handles
for community organizing, given these inequities, include public budgeting,
the use and abuse of public incentives, planning, and approvals by private

developers for large public/private projects that yield questionable benefits for neighborhoods. Legislation such as the Community Reinvestment Act
provides a more predictable path for raising these resource issues.

Community organizing/planning depends upon legislative or administrative guidelines (e.g., the CRA), the breadth and depth of community
coalitions, the supportiveness of public institutions, the profitability of
private development, and the localized nature of proposed solutions.
Organizing has been less successful when it has challenged private decision making and ownership or has advocated broad redistributional programs. Occasionally, community organizing has spilled over into political
campaigns that, when successful, have made bureaucracies more supportive of neighborhood economic development (Clavel and Wiewel, 1991).

What Does a Theory of Neighborhood
Economic Development Need?

The first section of this chapter identified two substantive requirements
for community economic development theory. First, the theory should
concern itself with the neighborhood economy and its linkages. It should
deal explicitly with economic variables and tell us something useful about
them. The concern is with levels of economic activity (variously measured), linkages, and economic outcomes for various groups.

A second substantive consideration mirrors the economic. The theory
must also be *sociopolitical* because the activity with which it is concerned
is rooted in a particular form of social institution and its practitioners act
explicitly within a political framework. This is a familiar aspect of community economic development, one that has shaped most conceptions of
the field and remains central to any theoretical approach to its problems.

The actual experience of practitioners suggests three additional, more
abstract requirements for a theory. First, the theory must be *instrumental*,
that is, it should provide guidance for action rather than simply a positive
explanation of how things behave. Second, it should also be *normative*, that
is, it must embody a set of objectives and arguments for their realization. This
assertion is closely related to the idea of instrumentality in the sense that an
instrumental theory demands some goal. It is also consistent with the entire
history of community economic development, which has been driven by the
desire for change and improvement in the conditions under which people live.

Third, a neighborhood economic development theory requires a *positive*
understanding of the contextual world in which the action that it deals with
will be played out. This term describes the kind of understanding of
relationships and interactions that is necessary for effective action.

Problems and Potentials in
Community Economic Development Theory

How well does current community economic development theory stand up to these criteria and how might it be improved? On the whole, it does not do well. While practice has generated an array of approaches and, to some extent, techniques, there is little in the way of an integrated theoretical framework that can give those elements coherence.

Although the variety of approaches suggests that this field is dominated by instrumental thinking, virtually none of the approaches contains the essential elements of instrumentality, namely, guidance as to when and how they may be more or less effective and predictive power about their impacts. The theoretical and substantive content of most books being written about economic development is thin (Mier and Fitzgerald, 1991). Careful assessment of the effectiveness of the practice elements laid out above has not been done. Not much is known about where they do and do not work or how they interact with each other. For instance, does providing job training cause an exodus of skilled people from the neighborhood rather than improvement in conditions? Good theory is essential as a guide to practice, both to build the arguments for innovation and to provide support to continue when innovations fail.

What about the normative content of current theory? Is that not an area of great strength? At one level, this is certainly the case. Practitioners are committed to community goals in the face of enormous obstacles and pressures, often to their own personal disadvantage. It is hard not to find an idealist even in the most battle-scarred exponent of neighborhood economic development. Yet, there are also problems of normative scope.

The nature of the field itself gives rise to serious conflicts in objectives among its proponents, probably far more so than in conventional economic development. Although practitioners are aware of such conflicts and live with them every day, virtually no theoretical attention has been paid to their nature and to their resolution. Where solidarity is conceived of as a supreme virtue, recognition and acceptance of the reality of conflicting objectives are difficult to attain. The relationship between activists outside government and their (former) friends within it during the mayoralty of Harold Washington in Chicago provides several, often painful examples (Brehm, 1991; Giloth, 1991). But when serious attention is paid to objectives and building consensus, as in the case of community organizing, the effort may be so great and the compromises so deep that the original economic objectives are lost in the political mobilization.

The third abstract characteristic of a community economic development theory—namely, an appropriate positive understanding of the contextual environment in which actions will be taken—presents a mixed picture. Certainly, economic development practitioners are all too aware of the constraints of the larger economy under which they work. Yet, it is also the case that community groups often do not understand or accept the reality of those constraints. In part, this is a tribute to their determination to achieve change in the face of what are formidable, indeed, impossible, obstacles. In this, however, they are quite similar to entrepreneurs who persist against great odds. For both groups, faith, hope, and determination are necessary bulwarks against failure in the face of a reality in which the objective probability of success is very low.

Admirable as it may be, this behavior can have serious consequences. For the single-minded entrepreneur, it may mean persisting beyond the point where failure is inevitable. For the community group, which has other objectives beyond development, it is more likely to give rise to a shift of focus away from economic goals toward others that appear to be viable and provide a sense of achievement.

Does this mean that community groups should know more about supply and demand? In some instances, yes; the history of worker takeovers demonstrates quite clearly the cost of illusions about business viability, and it is encouraging that more recent efforts are recognizing this most explicitly. But in a larger sense, conventional economic theory is not the only way in which to understand development. There is too much legitimate questioning and debate for that. The questions raised by conventional theory about the way in which markets work in a capitalist economy cannot, however, be ignored, no matter what injustice and distributional inequity may exist in their operation.

Similar problems exist in relation to the substantive content of community economic development theory. Clearly, a theoretical basis for this field must be both economic and sociopolitical in nature. Current practice revolves around activities that are economic in nature, but they are not integrated into an economic theory of community development that has much real content. In part, this is due to the history and evolution of the field, which has emerged from political struggles and attempts to preserve communities in the face of change and enhance their well-being in the face of exploitation, discrimination, and prejudice. A broadly accepted theory of community economic development is a long way off, and practice reflects this.

There are some elements upon which such a theory might be constructed, but there are also some formidable obstacles. Among the latter, perhaps none is more resistant than the concept of community itself. To set out the range of theoretical problems raised by this question is not possible in a short chapter. It is clear, however, that the idea of a theory of community development in which the fundamental unit of observation is so fuzzy and ill-defined must be subject to serious difficulties. Those who work and teach in this area tend to deal with the problem on an ad hoc basis. For a broader theoretical structure, however, that is inadequate.

Assuming that the concept of community can be managed satisfactorily, are there some substantive elements that might contribute to a theory of development and practice? One important consideration flows from the experience of the few instances, most notably in Chicago under Mayor Harold Washington, where community economic development has been brought to the forefront of policy. That experience suggests that the central role of community economic development action is to employ political means to achieve broadly redistributive goals through economic growth focused on particular populations and communities. The essential purpose is communitarian, but the method is to employ market mechanisms, reinforced by political efforts to generate the necessary resources for investment and to establish rules that are consistent with the larger market environment, yet supportive of community goals. There are, then, three necessary requirements to be satisfied by the process, and they must be mutually consistent for it to work.

First, the gains to communities should arise primarily from the enhancement of the productivity and creation of assets of community populations rather than solely through redistributive transfers. This implies that the realities of the market must be understood, recognized, and dealt with. Second, the objectives of community development are established at the local level, but they have to be realizable and consistent.

Third, the use of political power *in relation to community economic development* is primarily to facilitate the process of productivity enhancement and asset creation. The distinction here is between the politics of economic development in the community realm and the politics of other social goals, such as effective participation. It is, to some extent, an artificial distinction in practice, especially because the various elements of community development are intended to be mutually reinforcing, but from the perspective of economic development, it emphasizes the political willingness to come to terms with the market. This agenda is modest in comparison with the conceptions of sweeping social change that have historically informed some forms of community development, especially the socialist strand of thought. Nonetheless, it can form a viable basis for integrating actions for economic development in the 1990s.

A second consideration that might be brought to a theoretical structure focuses on what community economic development means and what the community itself brings to the process. It is important, and overdue, to sort out what it is in development that enhances the community and what it is that flows to the individual, the family, and the household or other smaller social units. This inevitably again raises the definitional problem of community. Do neighborhoods and communities in fact have collective or organic qualities for purposes of goal setting? The origins and character of community development suggest so, but this requires caution about the form and meaning of collective or aggregate goals. Despite many attacks from a variety of individualistic perspectives, a notion of community goals retains a strong hold on the imaginations and perceptions of people who participate in community development. From a developmental point of view, it leads to the issue of the relationship between the potential state that a community might realistically achieve and what it brings to the development process.

What does the community provide for the development process? While it is a common part of conventional practice to carry out some sort of resource assessment, there is little theoretical basis to assign meaning and significance to those resources for economic development. In practice, there also often is a reluctance to recognize resource deficits for fear of generating hostility. What exactly does constitute community or neighborhood resources? This is an issue that is worth much more theoretical exploration, especially with respect to those aspects of community that do not inhere in any particular individual or specific piece of property. For example, if a community is attractive for investment by virtue of its collective character or location, who should realize the gain and who should bear the cost? The traditional answers are increasingly rejected, not only in low-income or minority communities but wherever development is in prospect (Nyden and Wiewel, 1991). It is equally important, however, to ask what those collective resources actually are, how they affect development, and what will happen to them if development does occur.

References

Bolton, Roger (1992). "Place Prosperity vs People Prosperity Revisited: An Old Issue with a New Angle." *Urban Studies* 29 (2): 185-203.

Brehm, Robert (1991). "The City and the Neighborhoods: Was It Really a Two-Way Street?" In Pierre Clavel and Wim Wiewel (eds.), *Harold Washington and the Neighborhoods* (pp. 238-269). New Brunswick, NJ: Rutgers University Press.

Castells, Manuel (1983). *The City and the Grassroots.* Berkeley: University of California Press.

Center for Urban Economic Development (1987). *Community Economic Development Strategies: A Manual for Local Action.* Chicago: Center for Urban Economic Development, University of Illinois at Chicago.

Clavel, Pierre (1986). *The Progressive City: Planning and Participation, 1969-1984.* New Brunswick, NJ: Rutgers University Press.

Clavel, Pierre and Wim Wiewel (1991). *Harold Washington and the Neighborhoods: Progressive City Government in Chicago, 1983-1987.* New Brunswick, NJ: Rutgers University Press.

CWED (Chicago Workshop on Economic Development) (In preparation). [Social accounting study].

Downs, Anthony (1979). "Key Relationships Between Urban Development and Neighborhood Change." *Journal of the American Planning Association* 45 (4): 462-572.

Fainstein, Susan, Norman Fainstein, Richard C. Hill, Dennis Judd, and Michael P. Smith (1983). *Restructuring the City: The Political Economy of Urban Development.* New York: Longman.

Giloth, Robert (1988). "Community Economic Development: Strategies and Practices of the 1980s." *Economic Development Quarterly* 2 (4): 343-350.

Giloth, Robert (1991). "Making Policy with Communities: Research and Development in the Department of Economic Development." In Pierre Clavel and Wim Wiewel (eds.), *Harold Washington and the Neighborhoods* (pp. 100-120). New Brunswick, NJ: Rutgers University Press.

Giloth, Robert and John Betancur (1988). "Where Downtown Meets Neighborhood: Industrial Displacement in Chicago, 1978-1987." *Journal of the American Planning Association* 54 (3): 279-290.

Gunn, Christopher and Hazel Dayton Gunn (1991). *Reclaiming Capital: Democratic Initiatives and Community Development.* Ithaca, NY: Cornell University Press.

Harvey, David (1981). "The Urban Process Under Capitalism: A Framework for Analysis." In Michael Dear and Allan Scott (eds.), *Urbanization and Urban Planning in Capitalist Society* (pp. 91-122). London: Methuen.

Jencks, Christopher and Susan Mayer (1990). "The Social Consequences of Growing Up in a Poor Neighborhood: A Review." In M. McGeary and Larry Lynn (eds.), *Concentrated Urban Poverty in America.* Washington, DC: National Academy Press.

Kennedy, Marie and Chris Tilly (1987). "Secession and the Struggle for Community Control in Boston: A City Called Mandela." *North Star* (Spring): 12-18.

McKnight, John and Jody Kretzmann (1990). *Mapping Community Capacity.* Evanston, IL: Center for Urban Affairs and Policy Research, Northwestern University.

Mier, Robert and Joan Fitzgerald (1991). "Managing Economic Development." *Economic Development Quarterly* 5 (3): 268-279.

Mollenkopf, John (1983). *The Contested City.* Princeton, NJ: Princeton University Press.

Molotch, Harvey (1979). "Capital and Neighborhood in the United States: Some Conceptual Links." *Urban Affairs Quarterly* 14: 289-312.

NCCED (National Congress for Community Economic Development) (1991). *Changing the Odds: The Achievements of Community-Based Development Corporations.* Washington, DC: Author.

Nyden, Philip and Wim Wiewel (1991). *Challenging Uneven Development: An Urban Agenda for the 1990s.* New Brunswick, NJ: Rutgers University Press.

Ranney, David (1988). "Plant Closings and Corporate Disinvestment." *Journal of Planning Literature* 3(1): 22-35.

Ranney, David and John Betancur (1992). "Labor Force Based Development: A Community Oriented Approach to Targeting Job Training and Industrial Development." *Economic Development Quarterly* 6 (3): 286-296.

Scott, Allen (1980). *The Urban Land Nexus and the State.* London: Pion.

Teitz, Michael (1989). "Neighborhood Economics: Local Communities and Regional Markets." *Economic Development Quarterly* 3 (2): 111-122.

Walker, Richard (1978). "Two Sources of Uneven Development Under Advanced Capitalism: Spatial Differentiation and Capital Mobility." *Review of Radical Political Economics* 10 (3): 28-37.

Wellman, Barry (1976). *Urban Connections* (Research Paper No. 84). Toronto: Centre for Urban and Community Studies, University of Toronto.

Wiewel, Wim, Bridget Brown, and Marya Morris (1989). "The Linkage Between Regional and Neighborhood Development." *Economic Development Quarterly* 3 (2): 94-110.

Wiewel, Wim and Robert Giloth (1988). "Should Your Group Start a Business Venture?" In Mary O'Connell (ed.), *The Neighborhood Tool Box* (pp. 22-23). Chicago: Center for Neighborhood Technology.

Wiewel, Wim and Robert Mier (1986). "Enterprise Activities of Not-for-Profit Organizations: Surviving the New Federalism?" In Edward Bergman (ed.), *Local Economies in Transition: Policy Realities and Development Potentials* (pp. 205-225). Durham, NC: Duke University Press.

Wiewel, Wim and Nicholas Rieser (1989). "The Limits of Progressive Municipal Economic Development." *Community Development Journal* 24 (2): 111-120.

Wilson, William Julius (1980). *The Declining Significance of Race: Blacks and Changing American Institutions.* Chicago: University of Chicago Press.

Suggested Readings

Berger, Renee and Carol Steinbach (1992). *A Place in the Marketplace: Making Capitalism Work in Poor Communities.* Washington, DC: National Congress for Community Economic Development.

Castells, Manuel (1983). *The City and the Grassroots.* Berkeley: University of California Press.

Center for Neighborhood Technology (1986). *Working Neighborhoods: Taking Charge of Your Local Economy.* Chicago: Center for Neighborhood Technology.

Center for Urban Economic Development (1987). *Community Economic Development Strategies: A Manual for Local Action.* Chicago: Center for Urban Economic Development, University of Illinois at Chicago.

Clavel, Pierre (1986). *The Progressive City: Planning and Participation, 1969-1984.* New Brunswick, NJ: Rutgers University Press.

Clavel, Pierre and Wim Wiewel (1991). *Harold Washington and the Neighborhoods: Progressive City Government in Chicago, 1983-1987.* New Brunswick, NJ: Rutgers University Press.

Giloth, Robert (1988). "Community Economic Development: Strategies and Practices of the 1980s." *Economic Development Quarterly* 2 (4): 343-350.

Giloth, Robert, C. Orlebeke, J. Tickell, and P. Wright (1992). *Choices Ahead: CDCs and Real Estate Production in Chicago.* Chicago: Nathalie P. Voorhees Center for Neighborhood and Community Improvement, University of Illinois at Chicago.

Giloth, Robert and Wim Wiewel (1988). "Jobs: Should Your Group Start a Business Venture?" In Mary O'Connell (ed.), *The Neighborhood Tool Box* (pp. 22-23). Chicago: Center for Neighborhood Technology.

Harvey, David (1985). *The Urbanization of Capital: Studies in the History of Capitalist Urbanization.* Baltimore, MD: Johns Hopkins University Press.

Mollenkopf, John (1981). "Community and Accumulation." In Michael Dear and Allan Scott (eds.), *Urbanization and Urban Planning in Capitalist Society.* London: Methuen.

Molotch, Harvey (1979). "Capital and Neighborhood in the United States: Some Conceptual Links." *Urban Affairs Quarterly* 14: 289-312.

Nyden, Philip and Wim Wiewel (1991). *Challenging Uneven Development: An Urban Agenda for the 1990s.* New Brunswick, NJ: Rutgers University Press.

Teitz, Michael (1989). "Neighborhood Economics: Local Communities and Regional Markets." *Economic Development Quarterly* 3 (2): 111-122.

Vidal, Avis (1992). *Rebuilding Communities: A National Study of Urban Community Development Corporations.* New York: Community Development Research Center, New School for Social Research.

Wiewel, Wim, Bridget Brown, and Marya Morris (1989). "The Linkage Between Regional and Neighborhood Development." *Economic Development Quarterly* 3 (2): 94-110.

5

Ghetto Economic Development

WILLIAM WOODBRIDGE GOLDSMITH
LEWIS A. RANDOLPH

In the parking lot outside the church, a crowd of about 1,000 had gathered, a crowd that on any other night would have viewed [Mayor] Bradley with anything from affection to indifference. Professionals in suits, matrons holding their grandchildren, young men (none in gang regalia) from across the city, mainstream and fringe activists, nearly all of them black, had come to [the] AME [church] for what had been billed as a meeting to formulate a response to the now-infamous acquittals of the four policemen that had come down earlier that day in Simi Valley. The church was full even before the city officials arrived, so those outside listened to their own *ad hoc* collection of speakers and debated strategies.

Just a few blocks away, looting had already begun. "Don't do it in our community!" one speaker implored. Across the parking lot, people took up chants proposing alternative targets. "Beverly Hills!" shouted one cluster. "Parker Center" (LAPD headquarters) yelled another. (*In These Times* reporter, on the 1992 riots)

It was a hot August evening, and that small incident was all it took to ignite the frustrations, hostilities, and tensions that had been building up for years in Watts. A crowd gathered, and an ugly mood grew and erupted into violence. (Senator Fred Harris, on the 1965 riots)

I read the report . . . of the 1919 riot in Chicago, and it is as if I were reading the report of the investigating committee on the Harlem riot of '33, the report of the investigating committee on the Harlem riot of '43, the report of the McCone Commission on the Watts riot. (Psychologist Kenneth Clark, on the 1965 riots)[1]

Most ghetto residents are poor, some of them desperately so. They lack well-paying jobs and suffer from poor city services, underbudgeted schools, negligent welfare workers, and hostile, uncomprehending police. Even the 72,000 people who live in the 8th Ward of Washington, DC, in the Anacostia and Congress Heights neighborhoods, where some hold jobs in the federal bureaucracy, survive family incomes so low that the median is less than $17,000. That stands at about half the national level for whites and less than 42% of the median income for the Washington metropolitan area.[2] Of these families, 87% are renters, and 45% are headed by single women. Urban planner and George Washington University Professor Dorn McGrath says that to get "parity in the provision of municipal services" like garbage collection, sewer maintenance, and tree care, Anacostia had to sue the city (Chartrand, 1992).

People who live in better off neighborhoods and suburbs most often have higher paying jobs and usually enjoy adequate street cleaning and garbage collection, good schools, attentive public officials, and friendly, helpful police. Local businesses tend to invest profits locally. Although the 8th Ward is hardly a mile away from the Capitol, which lies just across the Anacostia river, it does not have a single locally owned restaurant, only outlets for fast-food chains that suck money out.[3]

The striking income inequalities and other gross disparities between ghettos, better off neighborhoods, and suburbs have led writers, policy-makers, and activists for years to theorize about the origins of ghetto poverty and to propose means for its reduction. Proposals "for improving conditions in black America—suburbanization, augmented employment, ghetto capitalism, [and] separatism—"continue to be made and continue to miss "the real issue, which is the creation of cohesive black political and economic power."[4] Yet similar proposals are once again being made, and once again they stop short of the real issues, arriving only at superficial substitutes for the deeper analysis and action that are required.

Three Fruitless Debates

I must again in candor say to members of this Commission it is a kind of Alice in Wonderland—with the same moving picture re-shown over and over again, the same analysis, the same recommendations, and the same inaction. (Kenneth Clark, in *The 1968 Report*, 1968: 483)

To illustrate how ghetto development proposals manage to miss the central issues, we focus on three debates. To make our own points, we take some liberty in paraphrasing, reorganizing, and simplifying the ideas of others, on all sides of the debates. Our point is neither to find merit nor to expose fault in the writings of others; this would be particularly inappropriate because we intentionally ignore many of these writers' subtle qualifications.

Our purpose is not to associate particular writers with specific theoretical or practical positions. Finally, it is not our purpose to pass judgment on one side or another of these particular debates. Instead, we aim to use many rich ideas from these writers to suggest that the debates themselves focus attention on the wrong issues. By doing this, we hope to expose flaws in commonly taken approaches to ghetto problems. We begin our discussion with the longest running debate, over the issue of Black Capitalism.

The Myth of Black Capitalism

Prominent national black leaders and political and social activists have since the Civil War proposed that, in the world's most advanced capitalist society, the best solution for the problems of ghetto residents would be more capitalism. In this society, where money talks and political changes follow the mighty dollar, no one should expect those who do not own capital to be able to claim profits. Nor should one hope people without businesses can offer to their neighbors the more general benefits of economic activity, like the in-group trickle-down of wages to area residents and local commercial purchases from home-owned business.

Many have argued the merits of Black Capitalism, differing with their critics over social, cultural, political, and economic issues that still shape and dominate discussions.[5] Although the main proposals are sometimes vague composites, they call for two essential advances: the capitalization of African American-owned firms, through transfers of funds, subsidies, and special programs for ghetto economic stimulus, and the broadening of general (mostly white) capitalist activity to provide more black people, mostly as employees but also as owners, with the fruits of the market.

In the late 1890s W. E. B. DuBois, the scholar, teacher, and socialist, debated Booker T. Washington, the black nationalist, conservative, capitalist. They disagreed initially over civil rights, Washington advocating "self help and racial upliftment," placing less emphasis on the pursuit of racial equality, and DuBois accusing Washington of advocating self-imposed segregation. They disagreed as well over the merits of replacing white with black control over the "black economy." DuBois was highly suspicious of programs that would create further divisions within the black community by encouraging exploitation by black capitalists (Marable, 1983: 140-167; Cruse, 1967: 330-335). Later, although DuBois acknowledged Marcus Garvey's back to Africa movement as a means of addressing the social, political, and economic problems confronting African Americans, he remained opposed to black capitalist exploitation of the black masses, even in the name of black nationalism.[6]

These disputes are relevant to U.S. ghettos today because many of the economic, political, cultural, and nationalist divisions of the era of DuBois and Washington remain in place. The Washington/DuBois disputes of the 1890s over economic advancement versus racial equality, the Garvey/DuBois disputes of the 1920s over separatism versus integration, and the conflict between black bourgeois politics and leadership for the black poor are still germane to the topic of ghetto development.

Many scholars argue in support of bringing more capitalism to the ghetto. Economists Walter Williams and Thomas Sowell stress the enormous losses in productivity and human capital that result from ghettos, laying blame on public regulation that limits the market (e.g., the high-priced Taxi Commission medallions that prevent poor people from legally driving taxis in New York City). They also oppose laws that encourage unions to exclude people from membership, enforcement of the prevailing and minimum wage requirements, and closed union shops. Sowell, in his focus on the harm done by antimarket regulations, asserts that race itself is not a significant factor in determining who gets left out (Sowell, 1984, 1987; Williams, 1982).[7]

Some scholars also at least implicitly support Black Capitalism because they fear government "handouts" that disempower the black community politically or economically. Williams, Sowell, and others have bitterly opposed the welfare state, affirmative action, and various Great Society programs. They are hostile toward welfare because they believe that it perpetuates a dependent relationship between blacks and the government, a dehumanizing experience, in which benefits flow only to bureaucrats (Williams calls them "poverty pimps"). The bureaucracy, according to Williams, Sowell, and other black conservatives, has no incentive to get people off welfare. For Sowell, just as for Booker T. Washington, capitalism is a more likely vehicle for "blacks to gain acceptance and upward mobility" (Marable, 1983: 174). If blacks would simply engage in self-help programs and work hard, and put less emphasis on achieving social equality, then whites would accept them over time.

Even prominent black nationalists have aligned themselves on some issues with the conservative and neoconservative camps, supporting market-oriented solutions to development problems. The list includes Roy Innis, former national director of the Congress on Racial Equality (CORE), Tony Brown, black media commentator, Floyd McKissick, former national director of CORE and founder of Soul City in North Carolina, and Robert Woodson, former Urban League official and currently the head of the Center for Neighborhood Studies, a housing group supported in part by the neoconservative American Enterprise Institute in Washington, DC.[8] This group has consistently supported conservative positions on social, political, and especially economic issues.

Innis and McKissick supported the concept of Black Capitalism (called separatist economics) during the 1960s and 1970s because they equated it with one of the most important cries for black power, that of self-determination. If African Americans somehow had control over the ghetto's economy, it is argued, they could then exercise some form of control over their communities, instead of having to rely consistently on white America for jobs and investment dollars. As Innis said to a national CORE convention in 1969:

> This separatist economics, as I choose to call it, is not essentially different from the basic principles of developmental economics employed by any people—for example, the Americans of the post-revolutionary period. It is the manipulation of the economy of black areas in a preferential way to obtain an edge and to protect the interests of the community; to place a membrane around the community that allows full commercial intercourse with outside business interests while setting pre-conditions and guidelines advantageous to the community for those who seek to operate within the community. This principle is known by many names, one of the more familiar being tariff.[9]

Although Innis introduced black *separatist* economics some 30 years ago and has since recanted the notion, he remains committed to market-oriented solutions. Tony Brown and Robert Woodson support Black Capitalism because like Sowell and Williams they also believe that economic power, not the overcoming of racism, is the key for advancement for African Americans. They embrace Black Capitalism because they believe liberal social programs have been an utter failure, contributing to welfare addiction. For them, the best cure for poor African Americans' addiction to government handouts is a swift kick of do-for-yourselfism (Marable, 1983: 176).

Some proponents of Black Capitalism have taken the argument further, asserting that until wealth is amassed and redistributed so that African Americans themselves can build political power and change laws and regulations, as well as make their own corporate practices, there would be no substantial improvements in the ghetto. In 1970 Richard America envisioned "a working mechanism . . . for the transfer of some major corporations to black control," a process involving transfer of eight major corporations in 15 years, a shift of about $2 billion a year, with control essentially in the hands of black stockholders. He cites precedent for such transfers in the railroad land appropriations of the nineteenth century, private rights to inventions sponsored with federal funds, and the Urban Renewal Program. The theory behind such a claim, of course, is entirely orthodox, positing if only implicitly that fair and free markets, and equal starting conditions, would have led on the average to equal achievements for blacks and whites. The oppression of slavery and the evils of discrimination imposed costs on blacks and advantages on whites that should now be redressed.[10]

Needless to say, corporations in the 1970s and 1980s did not act on these suggestions, nor have market forces or public subsidies shifted in favor of black capitalists. If anything, with the exaggerated polarization of U.S. income and wealth inequalities of the 1980s, we can reasonably suppose that the proportion and perhaps even the amount of capital held by African Americans is likely not to have increased. Census statistics released in October 1992 show an increased polarization of income among African Americans (in addition to a more unequal spread among all Americans). Unfortunately, this worsened black distribution is better evidence of poverty at the bottom than of capitalization at the top.

Nevertheless, examples exist of successful development efforts. A small, relatively successful effort at community economic development serves to bring this theoretical discussion down to a practical plane. The Edgemont Solar Garden project in Dayton, Ohio, was started by a group of concerned citizens in 1980. The project established a neighborhood garden on land that most residents considered to be the worst eyesore of this deeply impoverished black community. After a neighborhood coalition applied political pressure on city officials, they won a 20-year lease on the land to help residents "take control," in the words of a community leader, "of their community's economic destiny." The 90-plot community garden has employed 15 adults, seniors, probationers, and General Relief workers to produce and sell plants and fresh produce, African crafts, clothing, and other household items. The project involves training, community service, and construction of facilities, in addition to commerce. Because the neighborhood had no alternative access, no supermarket or produce store, the benefits are widespread.

In a way, this is a successful case of Black Capitalism. Notable, however, is that even this modest establishment of business activity in the neighborhood, which is only a small dose of Black Capitalism, succeeded because it benefited from a set of decidedly noncapitalist transfer payments. The land is leased from the city for only $1 a year; the federal Department of Housing and Urban Development (HUD) awarded a Neighborhood Demonstration Grant; and several nearby corporations and banks donated money (Lovelace, 1992). Without these funds, the business (which has cooperative characteristics as well) would never have been established.

Some, like Richard America, have as an assumption the notion that the American economy works just fine for whites; all blacks need is to be cut in on the game by means of a transfer of business ownership. The flaw in this reasoning is illustrated by its failure to account for numerous white victims of poorly regulated capitalist practices.[11] So, as appealing as this proposal is in terms of racial fairness, it sidesteps any discussion of the way inequality arises in market economies.

Others, like Thomas Sowell, propose to let capitalism run much more freely in the expectation that African Americans will benefit through its diffusion. This proposal is attractive in a very different way. Its strength comes from the assertion that African Americans will do as well as anyone else if inhibiting regulations are removed. The great weakness here is blindness to a powerful racism that still pervades many (white-dominated) economic institutions.[12]

The most damning criticism of the call for Black Capitalism is of course its patent impracticality. To begin with, there is nothing in any of the proposals that deals seriously with the kinds of constitutional, political, and social upheavals that would be needed for its rendering into public or private policy, much less its implementation. In fact, as a call for more of the same, that is, more private accumulation in the hands of a few, the proposal would seem to recommend strengthening precisely the forces that have brought about the problem. In any case, there is little to be done at the local level to shift capital away from whites and toward blacks in this way, and there is scant reason to expect any such shift at the national level.

The idea that Black Capitalism is a myth was developed and articulated by E. Franklin Frazier in the 1950s. Frazier maintained that black business was too insignificant to provide enough employment and income, that the black bourgeoisie was living in a world of make-believe:

> False ideas concerning the importance of Negro business have become a social myth . . . propagated among Negroes. [This is] one of the main elements in the world of "make-believe" which the black bourgeoisie has created to compensate for its feeling of inferiority in a white world dominated by business enterprise. (Frazier, [1957] 1969: 129)[13]

Frazier's analysis was expanded on by Earl Ofari in the late 1960s: Black Capitalism is a myth because it depends on the impossible—corporate America will not give up its economic control over the dependent ghetto (Ofari, 1970). Pan Africanist Stokely Carmichael (Kwame Ture) and political economist William Tabb essentially agree, and Robert Allen's analysis is very similar, too. Allen views Black Capitalism as just another ploy by corporate America to maintain its control over blacks in general and ghettos in particular (Allen, 1970: 2; Tabb, 1970: 21-34; Carmichael and Hamilton, 1967: 2-32; Hamilton, 1992). "The fact of black America as a semicolon, or what has been termed *domestic colonialism* . . . is . . . the most profound conclusion to be drawn from a survey of the black experience in America" (Allen, 1970: 2).

Indeed, the typical ghetto looks very much like a former colony, with extensive poverty, highly skewed incomes, high birth rates, and high infant mortality rates. There is little industry or commerce, and the area suffers from low labor productivity, low savings, and little investment. The major ghetto export is labor, which is tremendously underemployed, even at low wages. Almost all consumption is supplied from the outside, most of it conforming to external consumption standards, promoted by extensive advertising. Practically the entire ghetto economy is owned by outsiders (Goldsmith, 1974: 22-23).

These arguments against the idea of Black Capitalism as a main tool of development are hardly different in implication than those put forward by DuBois a hundred years ago, who said that African Americans need a cooperative economic scheme, an independent political party, and a cadre of militant leaders, essentially a social revolution to "promote maximum development of [the] community in toto" (quoted in Allen, 1970: 274-284).[14]

Poverty Culture and the Underclass

Theorists have borrowed a second approach to the problems of the ghetto directly from the literature on international economic growth. Here they argue that poor individual attitudes toward work, personal irresponsibility, and lack of enterprise are the main causes of underdevelopment. These unproductive attitudes result in large part from inappropriate cultural patterns in families and communities, including even poor practices of child rearing. These practices, seen as a "culture of poverty," condemn descendants to the poverty of their parents and their communities.

Beginning in the 1950s, when social scientists tried to find the key that would open the doors of the Third World to the benefits of modernization, several influential theorists posited in just this way that economic (under)development was ultimately caused by (in)correct community and family cultural practices. The right kinds of cultures, these theorists thought, created attitudes that promoted businesslike behaviors. Such entrepreneurial behavior, which would lead to economic success, was thought to be rooted in individual psychological orientations, personality traits that included orientation toward the future, self-denial, and hard work.[15]

After a while these contentious theories by highly respected development scholars percolated down to their colleagues in urban studies, who produced remarkably similar arguments about the problems of U.S. cities. These city scholars emphasized problems created by the poor values and bad attitudes held by ghetto dwellers, and they placed blame on the failures of liberal social policy in general. There were strong cultural, attitudinal, and value echoes as well in Daniel P. Moynihan's controversial 1965 report "The Negro Family."[16]

Edward Banfield in the 1960s went directly into the psychological realm of ghetto problems when he asserted that these problems arose from the manifold irresponsibility of ghetto residents, who he thought suffered great attitudinal problems. In this sort of theory, poor people in general, and ghetto young men in particular, are seen to be shortsighted, unenterprising, erotic, even mystical, and they "are said to suffer from the defects of their own (pathological) activity" (Goldsmith and Blakely, 1992: 4).

Charles Murray in the 1980s was more cautious about stereotypical explanations that laid the blame on deviant personal or family values, but he did blame liberal welfare policies of the 1960s and 1970s for creating a culture of dependency. His cultural focus bears strong similarity to the social-psychological work done in the international arena. He believes that reforms in the 1960s and 1970s in welfare, the criminal justice system, and schools promised too much that could not be delivered and thereby discouraged ghetto residents from working on their own for "slow incremental change" (Murray, 1985: 443-444).

Murray claims that Great Society job training programs, for example, led young men to expect they would get jobs with good pay, after which their unmet expectations prompted them to abandon further efforts. Similarly, in his view, increased welfare allotments and reduced stigma attached to the dole only served to make welfare more accessible and attractive than work. Murray feels that irresponsibility is also encouraged by permissiveness, in turn advanced by public policy. For example, the Supreme Count's *Miranda* decision limits police discretion in dealing with suspects, and tighter "restrictions on access to . . . juvenile arrest and court records" inhibit community condemnation (Murray, 1984: 69-84, 167-191).

This sort of conservative theory, applied to areas much broader than the question of ghetto economic development, became a staple of right-wing politicians by the 1980s: In a speech to the Catholic League for Religious and Civil Rights in October 1992, the U.S. attorney general worried that the country's moral fiber is weakened by rampant permissiveness. He opposed giving clean needles to addicts and distributing condoms in schools, and he advocated making abortions unconstitutional. In "contemporary moral philosophy," the attorney general said, simple desire acts "as an unreasoning tyrant over which reason, and therefore morality, has no influence" (Johnston, 1992).

Dan Quayle, with his tawdry focus on "family values" in the 1992 presidential campaign, caricatured even these cultural arguments, and a threatened Republican party used them for cynical advantage, in a replay of the 1988 Willie Horton scam. Spokespersons for the religious right in particular aimed their speeches and platform statements toward those eager to believe that ghetto problems are caused by the bad attitudes and poor values of ghetto dwellers. It seems fair to us to reject the Republican platform and Dan Quayle's rhetoric as simple-minded, reactionary, and racist, but the cultural-theory arguments by social scientists are considerably deeper and therefore deserving of more serious consideration. The latest version of these arguments has to do with the "underclass."

With the drastic increases of underemployment, poverty, and homelessness that began in the last half of the 1970s, a new term came back into popular use, the *underclass,* reminiscent of Karl Marx's "lumpen-proletariat." Investigations of the "underclass" multiplied and expanded rapidly in journalism and research institutes in the 1980s. It wasn't long before an entire industry grew up to analyze, enumerate, and (some would say) objectify these people. As one researcher puts it: "As the Reagan years progressed and panhandlers multiplied amid the bond salesmen, [opinion makers] became convinced that this underclass was growing. The idea that the black underclass was growing became especially common" (Jencks, 1991: 14). As a writer for *The New York Times* said in 1987:

> Social scientists have focused new energies on an "underclass" of Americans who live in near total isolation from mainstream society, and scholars are trying to learn more about the deteriorating inner-city areas where not working is the norm, crime is a commonplace and welfare is a way of life. (Wilkerson, quoted in Katz, 1989: 195)

Six years after Ken Auleta popularized the idea of the underclass in his 1981 *New Yorker* articles, William J. Wilson published an influential book, *The Truly Disadvantaged: The Inner City, the Underclass, and Public Policy.*[17] Wilson (like many others before him) aimed to show that externally imposed conditions (such as industrial decline, unemployment, lack of services, and selective suburbanization of middle-class blacks) had relegated a growing number of African Americans into underclass *communities,* ghettos in which they were not only isolated but attracted to pathological life-styles involving dependency on welfare, crime, and social disorganization. In much of the discussion and research that followed the naming of the underclass, larger hypotheses about social and economic reinforcement of ghetto isolation and misery have been lost, swamped by numerous smaller research questions concerning the size, location, and particular characteristics of segments of the population. Researchers ask,

for example, if the economic, or the moral, or the educational "underclass"
is growing or declining.[18]

Herbert Gans suggests caution against use of the word *underclass* because
it serves to isolate people, turn them into the undeserving poor, assuring the
rest of us that their concerns are not our fault. He finds *underclass* a danger-
ously convenient euphemism for harsher, old-fashioned terms like *paupers,
rabble, white trash,* and the *dangerous classes. Underclass* seems to him
the 1990s equivalent of the *culture of poverty* (Gans, 1990). Gans also
reminds us that, when Gunnar Myrdal used the term *underclass* in 1962,
his conception was both economic and relational: those in the underclass
were the chronically underemployed, people made poor and marginalized
by the changing structure of the economy, and Myrdal's focus was on the
need for economic reform.

Compared with the light criticism against those who use the term
underclass in this way, criticism of the culture-of-poverty theory has been
deep and broad. Critics from a variety of ideological positions have
explained the flaws in the even broader modernization theory, and they
have focused with special energy on its social-psychological variants,
including the culture of poverty. Those focusing on the international case
point out that poor people generally work hardest of all, often exhibiting
an extreme form of market rationality, that broad economic and political circumstances, perhaps accompanied by military or police repres-
sion, often limit the possibilities for organized improvement on the part
of poor neighborhoods, and that poor, underemployed, underhoused,
and publicly neglected workers frequently form the backbone of a
developing city's economy.[19]

Domestic critics, like Eliot Leibow in *Tally's Corner* (1967) or Jonathan
Kozol in *Rachel and Her Children* (1988), not to mention novelists and
essayists like Toni Morrison or James Baldwin, have demolished the false
notion that most inner-city poor people don't work hard, don't try to get
ahead. These writers provide ample evidence that on the contrary it takes
great effort and ingenuity for poor city people just to stay even, to cope
with a badly biased system.[20]

Historian Michael Katz begins *The Undeserving Poor* (1989) by pointing out that Americans may talk about poverty, but they leave some things
unsaid:

> Mainstream discourse about poverty, whether liberal or conservative, largely
> stays silent about politics, power, and equality. . . . Descriptions of the demog-
> raphy, behavior, or beliefs of subpopulations cannot explain the patterned
> inequalities. . . . These result from styles of dominance, the way power is
> exercised, and the politics of distribution.

This structural naïveté is the central problem with most work on both the underclass and the culture of poverty. The idea of the culture of poverty is a political problem, for it is used to justify conservative, victim blaming policies, offering poor people only "social work and therapy when they [need] economic justice and political mobilization." It is an empirical problem, because researchers have misidentified, mislabeled, and misinterpreted; and it is a theoretical problem, because it fails to account for the ways in which external social and especially economic forces condition individual and even group behavior.[21]

Rebuilding Versus Dispersal and Suburbanization

Shakespeare's words "to gild refined gold, to paint the lily, to throw a perfume on the violet" describe overkill, waste of effort, needless work. When economists John Kain and Joseph Persky delivered their "Gilded Ghetto" paper at the first research meeting of the newly formed Economic Development Administration, in early 1967, the assembled researchers and policymakers hooted and jeered. Kain and Persky had touched raw nerves.[22] By arguing (in the paper's title at least) that public monies intended to help impoverished residents of the ghetto might be spent most efficiently outside the ghetto, rather than inside, Kain and Persky inadvertently argued that things in the ghetto were OK, that any further worry would indeed be waste. Kain had written earlier that "the only efficient and satisfactory long run solution to ghetto problems [would be] suburbanization of the Negro population" (Kain, 1968: 20). Although Kain and Persky endorsed income transfers to meet the "immediate needs of ghetto residents," they wrote in favor of weakening "the ties" and "the geographic dominance of the ghetto." A central problem, they felt, was that the mere existence of the ghetto posed "serious implications . . . for the metropolis as a whole" and that many harmful problems affecting blacks directly were "dependent for much of their adverse impact on the very existence of the ghetto." They were joined in their support for decentralization by many other influential persons, among them urban researcher and writer Anthony Downs and NAACP labor secretary Herbert Hill (Goldsmith, 1974: 19; Kain and Persky, 1969: 3, 23).

Debates over questions regarding the location of relief, the bad influences of the ghetto on its residents, and especially the efficacy of ghetto economic development programs to stimulate business investment have continued ever since. A widely read early response was by Peter Labrie, in the *The Review of Black Political Economy*. Labrie argued that, where there existed heavy demand for unskilled labor, information about jobs would arrive from the suburbs to the ghetto, and although transportation to work from the ghetto to the suburbs would be costly in terms of time and money, this, too, was not an insurmountable problem (Labrie, 1970-1971). Many, who thought the spatial-mismatch hypothesis far too simple, claimed that job losses were caused mainly by generally high levels of unemployment and racial discrimination.

We might leave these "spatial-mismatch" debates to the historian, but those who propose policy still argue actively.[23] Researchers question what the most important location issues are, so as to decide whether public programs should support job growth in the ghetto, transportation access to the suburbs, or migration (from the ghettos to the suburbs). The debates about policy and causation circle around several familiar observations: Many jobs that require few skills and little education have disappeared, notably with the demise of traditional manufacturing industries; these job losses have been pronounced in central cities; ghetto residents previously depended on these industries for jobs; these residents lack access to suburban jobs for want of automobiles and adequate public transport; and in any case their skills and education do not meet the higher requirements of large numbers of jobs in the service and high-tech sectors, both in central cities and in suburbs (Goldsmith and Blakely, 1992: 132).

The central question in all this debate is quite simple: Does where you live have a major effect on your chance of finding out about, being offered, getting to, and holding a job? Specifically, how much is a potential employee harmed (in job search, initial employment, performance, and retention) by the fact that he or she lives in the ghetto. It turns out that the answers are complex. On the one hand, one sort of research into specific costs that can be attributed to ghetto residence comes up with inconclusive or even negative results. That is, some investigations show that, when people move from the ghetto to the suburbs, or when transportation access is expanded, job search performance improves only slightly. Another sort of research suggests that it matters a fair amount: Networks of information about jobs are extremely thin in ghettos, so ghetto residents don't find out about jobs; and potential employers have been found to discriminate against young job applicants simply because they fill out forms giving a "bad" address. As recently as 1991, William J. Wilson prominently reasserted his claim of "a growing mismatch between the location of employment and residence in the inner city," and his collaborators found evidence that living in the suburbs (other things being equal) enhanced the likelihood that a poor person would find a job (Wilson, 1991; Rosenbaum and Popkin, 1991).

The Political Economy of the Ghetto

Ghetto problems in the United States result from three kinds of forces: political hostility toward cities, deterioration of the national and especially the central-city economy, and institutional and personal acts of racism. Throughout the 1980s, an accelerated globalization of the economy, federal attention to international rather than domestic affairs, and an almost unbounded surge of personal selfishness set the stage. In these circumstances, the effects of political, economic, and racial forces have produced a near disaster for urban America in social, political, and moral terms. They have also been costly economically.

Politics

Ghetto (and broader central-city) economic problems were created, literally put in place, by 40 years of state and federal hostility to cities.[24] This mainly unconscious hostility took the form of pro-suburban, antiurban imbalances in subsidies for highways, unfair tax giveaways on housing, and a peculiar kind of fiscal relief for wealthy suburbs, exactly during and after the period when millions of African Americans migrated to the cities. Briefly: The Interstate Highway program provided federal funding to build 41,000 miles of freeways, probably the largest physical infrastructure project the world has ever known. This country's largest intervention into housing markets also came after World War II, in the form of Federal Housing Administration loan guarantees, which lowered the cost of borrowing money to build and buy homes, and in the form of an income-tax deduction for interest payments on home mortgages, a tax expenditure that now runs to a subsidy of about $80 billion a year (Goldsmith and Muhammad, 1992).

Together with expanding auto firms—and industries producing steel, rubber tires, and ancillary products—highways and houses gave form to what some call the American century, part and parcel of 40 years of economic boom. But the boom eventually left cities behind. The highway system not only built the suburbs, but it devastated the centers of cities, cutting giant gashes across them. The housing programs were restricted almost entirely to suburban, single family homes—and they were even restricted by law to segregated neighborhoods for white buyers. Insurance company leaders and mortgage bankers further pushed poor city neighborhoods down as they drew red lines and green lines on their maps to prevent the issuance of fire insurance or the granting of mortgages.

The coup de grace came with metropolitan Balkanization. It would be logical that the jurisdiction of city government expands as the city or metropolis grows, so that the beneficiaries of growth and adjustment pay its costs, but in the postwar boom exactly the opposite happened. Instead, political and fiscal jurisdictions multiplied and separated, so those families who moved out beyond the city's borders also escaped all responsibility for paying the bills. They used and ran down the public utility and highway and bridge systems, then they literally skipped town without paying their share into sinking funds for depreciation. People from the suburbs continued to use downtown cultural facilities, such as libraries, universities, and concert halls, as well as the central business districts, but they avoided taxation. It was a good deal for them: The suburbs boomed, their kids all went to college, and they took over the Congress. They forgot entirely about the costs. It is these costs with which cities—and of course the nation in the final analysis—are burdened today. As Kenneth Clark said of ghetto residents in 1965:

> This society knows . . . that if human beings are confined in ghetto compounds of our cities, and are subjected to criminally inferior education, pervasive economic and job discrimination, committed to houses unfit for human habitation, subjected to unspeakable conditions of municipal services . . . that such human beings are not likely to be responsive to appeals to be lawful, to be respectful, to be concerned with property of others. (*The 1968 Report*, 1968: 300)

Economics

In the 1960s and early 1970s, when great attention was last paid to the problems of ghetto economies, one main theme was echoed over and over again by the most prescient analysts: If ghetto conditions were to improve, there was a need for jobs, jobs, jobs. From across the political spectrum—Daniel Patrick Moynihan, Peter Labrie, Bennett Harrison, and Kenneth Clark, for example—there were reminders that, without job growth, satisfactory conditions in the ghettos would be impossible to achieve.

> Indeed, there seems to be no argument in the literature to counter the position that aggregate demand for black employment must be increased dramatically. . . . The only arguments are whether and how this is to be done. (Goldsmith, 1974: 20)

That was written in 1974, but it is just as true in 1993. The reverse is equally true: If jobs are lost, conditions will get worse. There are two main sources of jobs: the private economy and public employment. In the 1980s ghetto poverty increased, especially in the big industrial cities of the manufacturing belt. Why? "Because [this region] experienced massive industrial restructuring and loss of blue-collar jobs," a mainstay for black (and white) working-class incomes (Wilson, 1991: 465). Black unemployment is regularly about twice as high as white unemployment, and it increases about 2 percentage points every time white unemployment increases 1 point. Compounding employment declines in the private economy, public employment also shrunk, a catastrophe for African Americans, who have relied on public sector for job growth in the last decades.

Racism

Racial discrimination still plays an enormously negative role in the social experience of nearly every black American: It is an essential ingredient in the organization of schools, in the composition of neighborhoods, and in the arrangements of everyday life.

> Physically . . . white Americans leave the city . . . Psychologically, white Americans put walls up to the increasingly desperate plight of those, both black and white, who can't leave—those Americans . . . are stuck holding jobs in a Third World economy, establishing a sense of community in a desert where there is no water of hope.
>
> Now, it's not that there isn't racism. It's alive and well. It's not that police brutality doesn't exist. It does. It's not that police departments give residents a feeling of security. Few do. (Senator Bill Bradley, 1992)

For our analysis, the economic effects of racism matter most. Since about 1980 there has been much controversy in the social science literature about whether racism mattered any longer in the economy, but by the end of the decade researchers began producing incontrovertible findings. Even in business decisions regarding entry-level job offers—where Gary Becker's theories about competitive market forces promoting nondiscriminatory behavior should most strongly apply—racial prejudice plays a large part.

> Our interviews at Chicago-area businesses show that employers view inner-city workers, especially black men, as unstable, uncooperative, dishonest, and uneducated. Race is an important factor in hiring decisions. But it is not race alone: rather it is race in a complex interaction with employers' perceptions of class and space, or inner-city residence. (Kirschenman and Neckerman, 1991: 204)

For those black workers lucky enough to get jobs, the difficulties are not over. As Thomas Boston has shown, race-related earnings differentials result from discrimination at three stages of the labor market. At the final and most widely understood stage, minority workers on average get less pay for the same jobs. Even after accounting for differences in age, education, region, job experience, family size, and other factors, wages in many occupations are lower for African Americans than for whites. One stage earlier, and of more consequence, discrimination unfairly reduces incomes by limiting access of minority workers to preferred lines of industry, where jobs are better and higher paying. Worse yet, at the first state of labor allocation, inside industrial sectors, African Americans are "disproportionately concentrated among low-paying occupations" even after controlling "for job-related attributes, age and other demographic differences" (Boston, 1988: 73-87).

Alternatives

The Negro problem looked at in one way is but the old world questions of ignorance, poverty, crime, and the dislike of the stranger. On the other hand it is a mistake to think that attacking each of these questions single-handed without reference to the others will settle the matter: a combination of social problems is far more than a matter of mere addition—the combination itself is a problem. (DuBois [1899], 1967: 385)

Are there alternatives to these antiurban, antiminority politics and economics? Could better U.S. policy result not only in rehabilitated cities, more equal income distribution, and less poverty in the cities and elsewhere but also in a reconstructed economy and society, which would compete more effectively overseas? We believe such policy is possible, but we are not optimistic about its adoption.[25]

If a community of poverty produces illness, poor education, negative attitudes, poor work habits, and low productivity, it is sure eventually to cause problems for the country as a whole. Inasmuch as urban poverty is closely associated with collapsing and devalued housing, streets, bridges, utility systems, and other components of the infrastructure, it also symbolizes (and results from) a huge drag on productivity and a restraint on industrial flexibility nationally. These problems go far beyond urban poverty, but they are closely connected to it.

There can be no doubt that enormous improvements are required in federal legislation and huge increases in federal budgets for cities. New and expanded programs are required above all in industrial policy to generate employment; in education, to provide skills and the potential for human growth; in welfare, to emphasize meeting needs and rewarding enterprise; and in health care, to provide universal services and control costs. Unless these programs are forthcoming for the nation, prospects for improvements for cities and ghetto populations are meager.

Meanwhile, cities and regions will be beset with problems, and persons promoting local economic development must push up a steep slope. Just as theories about local development are constrained and conditioned by national and international considerations, so practice in local economics must accommodate to larger and more general forces. Recession or slow growth, globalization and increased competition from cheap labor, and continual pressure from reactionary political forces will hamper most efforts.

Ghetto residents and various allies will, however, continue to press for change. Our review suggests that their practical efforts should pay close attention to politics, economics, and race. That is, their efforts, as a subset of local economic development, should stress the connections between political empowerment, economic expansion, and the fight against racial discrimination.

Notes

1. Commentators have employed similar language to report on racism-related riots in the United States throughout the twentieth century. The quotations are from a *In These Times* reporter on the 1992 events, Fred Harris on the 1965 events, and Kenneth Clark on events in four different years (Meyerson, 1992; Harris and Wilkins, 1988: 5; *The 1968 Report of the National Advisory Commission*, 1968: 483)

2. The 1989 median income in the Washington, D.C., metropolitan area was $40,821 for families and $43,942 for households.

3. See Gunn and Gunn (1991) for discussion of the economic drain on communities by fast-food joints and other businesses.

4. What one author of this chapter observed two decades ago remains unchanged (Goldsmith, 1974: 17).

5. In what follows, we have used extensively the analysis of Manning Marable (1983). See also the work of Cruse (1967) and Allen (1970).

6. Their final debate involved politics: The black middle class failed to provide effective leadership because (according to Marable) this group has generally been "either clients of larger corporate interests, or they excelled in the electoral game for profit and ego gratification" (Marable, 1983: 146-147; also see DuBois, 1940: 276-278; Cruse, 1967: 328-336; Boston, 1988).

7. Sowell also asserts (incorrectly, we believe) that West Indian economic success in the United States contradicts arguments that racism harms blacks economically. See note 13 below.

8. All these persons opposed groups seeking socialist alternatives for reconstructing America's ghettos (Marable, 1983: 174-175).

9. Pugh and Haddad (1969) provided a primer on Black Capitalism of the period, citing Innis on page 58.

10. For further discussion of Richard America and related ideas of Black Capitalism, see Goldsmith (1974: 21). A much smaller reparations claim has been made successfully (however

inadequately) by Japanese Americans in the western states against the costs of their internment in World War II; others have been made by Native Americans. Our focus on the African American condition reflects our conviction that Black ghetto problems are very serious, but we do not mean to minimize the often similar problems of similarly confined residential groups, such as impoverished Latinos in cities or Native Americans on reservations.

11. Although it is true that blacks are proportionately much worse off, nevertheless, nearly two-thirds of America's poor families are white (Goldsmith and Blakely, 1992: Figure 2.9, 32).

12. This argument has been elaborated and documented recently (Bell, 1992).

13. As it turns out, there is something more to the story. Boston (1988) shows that race-based capital accumulation has actually become not more but less of a possibility with successive decades. In the not too distant past, many African American neighborhoods were graced with (or dominated by) numerous locally based small businesses whose owners and in some cases managers constituted a local petit bourgeoisie, maintained middle-class patterns, and exercised public leadership. Later, the growth of corporate capital and the destruction of neighborhoods made many of these small businesses uneconomical, and the successors of small black business owners turned into managers and technicians, in jobs outside the ghetto, minor cogs in the machinery of much larger white corporations.

14. For a fuller treatment, see DuBois's original, *Dusk of Dawn* (DuBois, 1940).

15. The scholars who reported on research affirming these associations included anthropologist Oscar Lewis, who studied modernization and the culture of poverty in Mexico and who coined the phrase "culture of poverty" (Lewis, 1961, 1965), economist Everett Hagen, who looked at child rearing practices in the city of Medellin and other parts of Antioquia, Colombia (Hagen, 1968), and political scientist Daniel Lerner, who studied psychological and social change in the city of Ankara and its surroundings in Turkey (Lerner, 1967). Somewhat later, psychologist David McClelland found millennia of time-lagged correlations between good psychological attitudes (a need-for-achievement, called "N-ach," indicated by patterns on pottery chards) and bursts of economic development (measured by energy source depletion). He eventually took his fascinating research findings and applied them in naive ways to U.S. foreign aid programs for the training of new entrepreneurs in India and elsewhere (McClelland, 1961, 1969).

16. Edward Banfield wrote *The Unheavenly City* in 1968; George Gilder wrote *Wealth and Poverty* in 1981; Charles Murray wrote *Losing Ground* in 1984; and Thomas Sowell wrote *A Conflict of Visions* in 1987. The Moynihan report was published in 1965.

17. Katz notes that *Time* devoted a cover story to the underclass in 1977. In academia, the Social Science Research Council sponsors a program to develop scholarship on the underclass (Jencks and Peterson, 1991; see the literature review by Blakely, 1992).

18. Jencks examines five underclasses: jobless men, jobless women, the undereducated, the violent, and the "reproductive" underclass (Jencks, 1991).

19. See the ample and convincing discussions in Valentine (1968), Leacock (1971), Perlman (1976), and Young (1971).

20. This literature has a long and continuing history (Whyte, [1943] 1969; Leibow, 1967; Kozol, 1988; Katz, 1989; Kotlowitz, 1991).

21. Both quotations are from Katz (1989: 7, 39), who suggests the political, empirical, theoretical critique.

22. In their own defense, they had to spend conference time bearing witness to their commitment to the civil rights struggle.

23. The term *spatial-mismatch* was coined by *The 1968 Report of the National Advisory Commission* (1968); Kain (1993); see Wilson (1991).

24. Specific types of hostility and the "acts" by which hostility was implemented have been documented by a number of journalists and scholars (Gottdiener, 1985; Schmandt, 1988).

25. For a discussion of local versus federal political possibilities, see Goldsmith and Blakely (1992: chap. 5). The 10-page document from the 1992 presidential campaign, "A New Covenant for America's Cities," put out by the Clinton for President Committee, focuses on education and housing and mentions infrastructure and welfare, but, when it gets to the economy, it is vague, suggesting more transfers, better tax treatment, and training programs. For reviews of progressive reforms in cities, see Clavel (1986), Clavel and Weivel (1991), and Krumholz and Forester (1990).

References

Allen, Robert L. (1970). *Black Awakening in Capitalist America.* New York: Anchor.

America, Richard (1970). "What Do You People Want?" *Review of Black Political Economy* 6 (Spring-Summer): 45-57.

Banfield, Edward C. (1968). *The Unheavenly City.* Boston: Little, Brown.

Bell, Derrick (1992). *Faces at the Bottom of the Well: The Permanence of Racism.* New York: Basic Books.

Blakely, Edward J. (1992). "Villains and Victims: Poverty and Public Policy." *Journal of the American Planning Association* (Spring): 248 ff.

Boston, Thomas D. (1988). *Race, Class and Conservatism.* Boston: Unwin Hyman.

Bradley, Bill (1992). "A Plea for Racial Dialogue: A Plan to Save Our Cities." *In These Times* (Special Edition) 16 (24; May 13-19): 18-19.

Carmichael, Stokely and Charles V. Hamilton (1967). *Black Power.* New York: Vintage.

Chartrand, Sabra (1992). "A Washington District That's a World Apart." *The New York Times* (October 8): A20.

Clavel, Pierre (1986). *The Progressive City.* New Brunswick, NJ: Rutgers University Press.

Clavel, Pierre and Wim Weivel (eds.) (1991). *Harold Washington and the Neighborhoods.* New Brunswick, NJ: Rutgers University Press.

Clinton for President Committee (1992). "A New Covenant for America's Cities" (Photocopy).

Cruse, Harold (1967). *The Crisis of the Negro Intellectual.* New York: Quill.

DuBois, W. E. B. [1899] (1967). *The Philadelphia Negro.* New York: Schocken.

DuBois, W. E. B. (1940). *Dusk of Dawn.* New York: Harcourt, Brace.

Frazier, E. Franklin [1957] (1969). *Black Bourgeoisie.* London: Collier.

Gans, Herbert (1990). "Deconstructing the Underclass: The Term's Dangers as a Planning Concept." *Journal of the American Planning Association* (Summer): 271-277.

Gilder, George (1981). *Wealth and Poverty.* New York: Basic Books.

Goldsmith, William W. (1974). "The Ghetto as a Resource for Black America." *The Journal of the American Institute of Planners* 40(January): 17-30.

Goldsmith, William W. and Edward J. Blakely (1992). *Separate Societies: Poverty and Inequality in U.S. Cities.* Philadelphia: Temple University Press.

Goldsmith, William W. and Maryam Muhammad (1992). "City Limits: Lessons from L.A." *The Bookpress* (September): 1, 15, 20.

Gottdiener, M. (1985). "What Happened to Urban Crisis?" *Urban Affairs Quarterly* 20 (June): 421.

Gunn, Christopher and Hazel Dayton Gunn (1991). *Reclaiming Capital: Democratic Initiatives and Community Development.* Ithaca, NY: Cornell University Press.

Hagen, Everett (1968). *The Theory of Economic Development.* Homewood, IL: Irwin.

Hamilton, Dr. Charles V. (1992). Interviewed by Randolph (April 5).

Harris, Fred R. and Roger W. Wilkins (eds.) (1988). *Quiet Riots: Race and Poverty in the United States.* New York: Pantheon.

Innis, Roy (1969). "Separatist Economics: A New Social Contract" in William Haddad and
    G. Douglas Pugh (eds.), *Black Economic Development* (pp. 50-59). Englewood Cliffs,
    NJ: Prentice-Hall.
Jencks, Christopher (1991). "Is the American Underclass Growing?" in Christopher Jencks
    and Paul E. Peterson (eds.), *The Urban Underclass* (pp. 28-100). Washington, DC:
    Brookings Institution.
Jencks, Christopher and Paul E. Peterson (eds.). (1991). *The Urban Underclass.* Washington,
    DC: Brookings Institution.
Johnston, David (1992). "Justice Official Sees Weakening of Moral Fiber." *The New York
    Times* (October 8).
Kain, John F. (1968). "Housing Segregation, Negro Employment, and Metropolitan Decen-
    tralization." *Quarterly Journal of Economics* 37 (May): 175-197.
Kain, John F. (1993). "The Spatial Mismatch Hypothesis: Three Decades Later." *Housing
    Policy Debate* 3: 371-460.
Kain, John F. and Joseph Persky (1969). "Alternatives to the Gilded Ghetto." *Public Interest*
    14 (Winter): 74-87.
Katz, Michael B. (1989). *The Undeserving Poor: From the War on Poverty to the War on
    Welfare.* New York: Pantheon.
Kotlowitz, Alex (1991). *There Are No Children Here: The Story of Two Boys Growing Up in
    the Other America.* New York: Doubleday.
Kozol, Jonathan (1988). *Rachel and Her Children.* New York: Crown.
Krumholz, Norman and John Forester (1990). *Making Equity Planning Work: Leadership in
    the Public Sector.* Philadelphia: Temple University Press.
Labrie, Peter (1970-1971). "Black Central Cities: Dispersal or Rebuilding" (Parts I and II).
    The Review of Black Political Economy 1 (Autumn, 1970): 1-20; 2 (Winter-Spring,
    1971): 78-99.
Leacock, Eleanor (1971). *The Culture of Poverty: A Critique.* New York: Simon & Schuster.
Leibow, Elliott (1967). *Tally's Corner: A Study of Negro Street Corner Men.* Boston: Little,
    Brown.
Lerner, Daniel (1967). *Communication and Change in the Developing Countries.* Honolulu:
    East-West Center Press.
Lewis, Oscar (1961). *The Children of Sanchez.* New York: Random House.
Lewis, Oscar (1965). *La Vida.* New York: Random House.
Lovelace, Dean (Director of Neighborhood Development, Office of Strategies for Develop-
    ment, University of Dayton) (1992). Interviewed by Randolph (May 19, July 17).
Marable, Manning (1983). *How Capitalism Underdeveloped Black America.* Boston: South
    End.
McClelland, David C. (1961). *The Achieving Society.* Princeton, NJ: Van Nostrand Reinhold.
McClelland, David C. (1969). *Motivating Economic Achievement.* New York: Free Press.
Meyerson, Harold (1992). "Casualties of the Los Angeles Riot." *In These Times* (Special
    Edition) 16 (24; May 13-19): 2.
Moynihan, Daniel P. (1965). "The Negro Family: The Case for National Action." Washington,
    DC: U.S. Department of Labor (March).
Murray, Charles (1984). *Losing Ground.* New York: Basic Books.
Murray, Charles (1985). "Have the Poor Been Losing Ground?" *Political Science Quarterly*
    100 (1, Fall): 427-445.
Myrdal, Gunner (1962). *Challenge to Affluence.* New York: Pantheon.
Ofari, Earl (1970). *The Myth of Black Capitalism.* New York: Monthly Review Press.

Perlman, Janice (1976). *The Myth of Marginality: Urban Poverty and Politics in Rio de Janeiro.* Berkeley: University of California Press.

Pugh, G. Douglas and William F. Haddad (eds.) (1969). *Black Economic Development.* Englewood Cliffs, NJ: Prentice-Hall.

Rosenbaum, James E. and Susan J. Popkin (1991). " 'We'd Love to Hire Them, But . . .': The Meaning of Race for Employers" in Christopher Jencks and Paul E. Peterson (eds.), *The Urban Underclass* (pp. 203-235). Washington, DC: Brookings Institution.

Schmandt, Henry J. (1988). "Urban Research 1965-1987: A Content Analysis of Urban Affairs Quarterly." *Urban Affairs Quarterly* 24 (1, September): 3-33.

Sowell, Thomas (1984). *Civil Rights: Rhetoric or Reality?* New York: Morrow.

Sowell, Thomas (1987). *A Conflict of Visions.* New York: Morrow.

Tabb, William K. (1970). *The Political Economy of the Black Ghetto.* New York: Norton.

The 1968 Report of the National Advisory Commission on Civil Disorders (1968). New York: New York Times Co.

Valentine, Charles (1968). *Culture and Poverty.* Chicago: University of Chicago Press.

Whyte, William Foote [1943] (1969). *Street Corner Society.* Chicago: University of Chicago Press.

Williams, Walter (1982). *The State Against Blacks.* New York: Free Press.

Wilson, William Julius (1987). *The Truly Disadvantaged: The Inner City, the Underclass, and Public Policy.* Chicago: University of Chicago Press.

Wilson, William Julius (1991). "Public Policy Research and the Truly Disadvantaged" in Christopher Jencks and Paul E. Peterson (eds.), *The Urban Underclass* (pp. 460-481). Washington, DC: Brookings Institution.

Young, Frank (1971). "A Macroscopic Interpretation of Entrepreneurship" in Peter Kilby (ed.), *Entrepreneurship and Economic Development.* New York: Free Press.

Suggested Readings

Allen, Robert L. (1970). *Black Awakening in Capitalist America.* New York: Anchor.

Auletta, Ken (1982). *The Underclass.* New York: Random House.

Baldwin, James (1974). *If Beale Street Could Talk.* New York: Dial.

Bell, Derrick (1992). *Faces at the Bottom of the Well: The Permanence of Racism.* New York: Basic Books.

Boston, Thomas D. (1988). *Race, Class and Conservatism.* Boston: Unwin Hyman.

Clavel, Pierre and Wim Weivel (eds.) (1991). *Harold Washington and the Neighborhoods.* New Brunswick, NJ: Rutgers University Press.

Ellwood, David (1988). *Poor Support: Poverty in the American Family.* New York: Basic Books.

Glasgow, Douglas G. (1981). *The Black Underclass.* New York: Vintage.

Goldsmith, William W. and Edward J. Blakely (1992). *Separate Societies: Poverty and Inequality in U.S. Cities.* Philadelphia: Temple University Press.

Harrison, Bennett (1974). "Ghetto Economic Development: A Survey." *Journal of Economic Literature* 12(1): 1-37.

Katz, Michael B. (1989). *The Undeserving Poor: From the War on Poverty to the War on Welfare.* New York: Pantheon.

Krumholz, Norman and John Forester (1990). *Making Equity Planning Work: Leadership in the Public Sector.* Philadelphia: Temple University Press.

Marable, Manning (1991). *The Crisis of Color and Democracy: Essays on Race, Class, and Power.* Monroe, ME: Common Courage Press.

Massey, Douglas S. (1992). "American Apartheid: Segregation and the Making of the Underclass." *Poverty and Race* 1 (September).

Morrison, Toni (1992). *Racing Justice and Engendering Power.* New York: Pantheon.

Randolph, Lewis A. (1990). "Development Policy in Four Cities." Unpublished dissertation, Department of Political Science, Ohio State University.

Sowell, Thomas (1981). *Markets and Minorities.* New York: Basic Books.

Wilkerson, Isabel (1987). "New Studies Zeroing in on Poorest of the Poor." *The New York Times* (December 20), A26.

Wilson, William Julius (1987). *The Truly Disadvantaged: The Inner City, the Underclass, and Public Policy.* Chicago: University of Chicago Press.

PART III

Labor and Capital Theories

6

Labor Force, Education, and Work

JOAN FITZGERALD

As the field of economic development evolves, numerous debates over what activities it includes and who its practitioners are have emerged. Historically, local economic development practice has been dominated by efforts to maintain an area's economic base by attracting business, including creating the conditions commonly referred to as a "good business climate." But almost from the beginning, there has been a movement to broaden this conception of practice to include goals of "advocacy planning," "progressive planning," and "equity planning" and to broaden the base of actors involved in economic development.[1] In many cases, these efforts to reshape practice are associated with specific events, such as the election of a new mayor or council or the decline of a particular industry.[2] Out of these efforts emerge new ways of viewing economic development practice.

We are now in a reshaping period. Several recent, highly publicized reports argue that the economic competitiveness of the United States is jeopardized by a mismatch between jobs of the future and the skills of the population. Almost weekly new reports surface detailing the failure of the U.S. education system at all levels. Consensus has emerged on many fronts that economic development must be better linked to education and training. We are not clear, however, about what this linkage means in practice. Clearly, there is no theory that drives the reshaping.

Ultimately, how we link education, training, and economic development cannot be decided outside of the equity and inclusion debates that have been ongoing among practitioners and academics in the field. Yet, while we debate what we should be doing, initiatives to make education more responsive to the economy already are under way in most states and in hundreds of cities.

This chapter examines several assumptions that guide these reforms and reflects on their promises and problems in establishing the appropriate links between economic development, education, and training. The first question focuses on defining the problem. While most reform efforts are based on the premise that a skills mismatch does exist, there is plenty of evidence to suggest that it may be overstated. Further, the skills mismatch is complexly interwoven with race issues in most cities. The second question focuses on who the appropriate actors in the new linkage and reform efforts should be. Specifically, it examines the role of business, educators, labor, and community organizations as change agents. It is argued that the actors involved are critical in shaping the focus of the reform initiatives and thus the programmatic outcomes.

The Skills Mismatch Debate

Two recent and highly publicized reports—*Workforce 2000*, the Hudson Institute's Report to the Department of Labor, and *America's Choice: High Skills or Low Wages*, published by the National Commission on Education and the Economy—are frequently cited as evidence of a serious skills mismatch that is preventing U.S. firms from competing economically with countries such as Germany and Japan. In fact, the two reports are based on two strikingly different assumptions about the skill requirements of jobs of the future (Johnston and Packer, 1987; National Center on Education, 1990).

Workforce 2000 starts with the assumption that the U.S. economy's structural shift from manufacturing to services has created the demand for more highly skilled workers. Following Bell (1973), the study assumes that increasing skill requirements are a function of the shift to a postindustrial economy. In contrast, the *America's Choice* study found most employers did not see a need for more work force training in the future. The reason is that, rather than investing in skills upgrading for their workers, many employers have responded to intensified competition by making the low-wage, low-skill choice of automating as well as substituting cheap labor over product and process innovation and human capital investment (also see Dertouzos et al., 1989). This finding suggests that the competitive high-skill economy of the future, though desirable, is not here yet.

Two trends are at issue. The first is how structural shifts in the economy are changing the skill requirements of jobs. The second is how technology is being employed in the workplace to change skill requirements within occupations. Both analyses are necessary to understanding skill requirements of jobs of the future.

Mishel and Teixeira (1991) also analyzed the same U.S. Bureau of Labor Statistics data used in the *Workforce 2000* report and found little evidence to suggest that structural shifts in the economy will significantly change the occupational composition of employment. The difference in their findings stems from their including all jobs, while *Workforce 2000* focuses only on new jobs. While some new high-skill jobs will be created, their findings show these jobs will be more than counterbalanced by a rise in low-skill jobs. Further, Mishel and Teixeira indicate that low-skill service sector jobs will be the largest contributors to employment growth between 1984 and 2000.

Nor is there conclusive evidence that new technologies are increasing the skill requirements within many occupations. Reviews of the skills mismatch literature discuss how varying methodological approaches have led to conflicting conclusions (see Form et al., 1988; Bailey, 1991). Case studies tend to predict an overall deskilling of jobs. Following Braverman (1974), several industry studies have revealed how technology is intro-duced to increase control of management over workers and to raise the intensity of work. For example, case studies of the printing (Wallace and Kalleberg, 1982), metalworking (Cann et al., 1991), aerospace, agricultural implement, and auto industries (Shaiken et al., 1986) demonstrate how the introduction of new technologies reduced worker autonomy and control over the production process. These findings suggest that companies implement new technologies in ways that reduce the need to invest in worker training. It is difficult to generalize from case studies, however, because they tend to focus on frontline manufacturing workers, and not white-collar jobs, and on large, rather than smaller, industries (Form et al., 1988).

The mixed evidence suggests that the effect of technology on skill requirements and the organization of production are contingent on other sectoral and organizational factors (Spenner, 1985). For example, skills upgrading is more likely to occur with the introduction of new production technologies in small, less bureaucratized firms (Kelley, 1990). Indeed, Hayes and Jaikumar (1988: 81-82) conclude that many highly bureaucra-tized firms hesitate to introduce new technologies because it will destabi-lize the chain of command within the firm:

> The interfunctionality engendered by the new manufacturing can mean much more informal cooperation at low levels in the organization—between engi-neers and market analysts, designers and manufacturers. This kind of teamwork is unnatural behavior for companies whose structures, staffing policies, and performance measures operate according to command-and-control mentality.

The effect of technology on skill requirements also may vary by occu-pation within a firm. In the automobile industry, a small number of skilled workers experience skills upgrading with the introduction of new technol-ogy, but for the majority of frontline workers, skills downgrading has been the result (Roberts, 1984; Milkman and Pullman, 1991).

The picture indeed is confusing. Technology can be introduced as a means of exerting more control over workers and can be withheld for the same reason. Technology can be used to make work more challenging or to routinize tasks. As suggested by the NCEE report, the critical word is choice. Technology is not a force in itself that drives the skill requirements of jobs upward. Rather, firms choose how to use technology in organizing production. Unfortunately, the NCEE report confirms that most U.S. firms are not making the high-wage, high-skill choice (also see U.S. Congress, 1990). Firms have more incentive to organize production in ways that reduce labor costs: the low-skill choice (Rumberger, 1987). Even in firms making considerable investment in modernizing machinery, there is seldom comparable investment in training or reorganization of production (Rogers and Streeck, 1991).

Training as Economic Development

For over a decade theories of training have been built on notions of work skills as "human capital." The preceding discussion makes amply clear the simplicity and limitations of that theoretical approach. These conclusions suggest that addressing the skills mismatch problem requires both demand-side efforts to improve technology use and supply-side efforts to equip workers with the skills needed to use new production technologies. Addressing the demand side is familiar territory for economic development practice. Technology transfer and use programs increasingly are becoming a part of state and local economic development activity.[3]

In addition to technology transfer, economic development programs also offer training support. The Ohio Department of Development's Industrial Training Program (OITP) is typical of such efforts. OITP was created in 1980 to provide financial assistance to companies in training new or existing workers when new technologies are incorporated. The program has invested $70 million in training for 500,000 Ohio workers. In such programs, training is a means to the end of keeping firms in the state competitive and thus profitable.

The focus of the OITP and similar state programs throughout the country on providing training to upgrade the skills of current workers is justified, given that 85% of the work force at the turn of the century is already working now (Salamon, 1989). Yet training efforts need to be expanded to include those not in the work force. The biggest effect of technology will not be to displace existing workers but to exclude those with low-skills who are not working (Cyert and Mowery, 1987; Choate and Unger, 1986). Increasingly, these categories are dominated by inner-city minorities (Hill and Rock, 1989; Goldsmith and Blakely, 1991). The lower educational attainment of blacks relative to whites is a contributing factor in their higher rates of unemployment (Cotton, 1985; Farley, 1987; Moore and

Laramore, 1990). While spatial mismatch and discrimination are equally significant barriers to employment and opportunity, higher levels of educational attainment are still necessary for improving employment opportunity for blacks and other minorities.[4]

Getting those with the lowest skills to the level of employability will require intensive and comprehensive solutions. Economic development practice has been reluctant to support such efforts. Again, the Ohio case illustrates this tendency. In 1986 a minority set-aside program was established under the OITP, with the goal of increasing the work skills of economically disadvantaged, chronically unemployed black males. The High Unemployment Population Program (HUPP) places participants in firms in paid on-the-job training positions while in associate degree programs or apprenticeship training.[5] The program pays student tuition, provides support services such as child care and transportation, and reimburses firms for replacement employees while students are in the classroom.

This set-aside program was met with considerable resistance from DOD Economic Development Specialists within the OITP system. For example, although the Director of DOD took personal interest in pulling together a steering committee to establish program guidelines, without pressure from middle management this activity proceeded at a snail's pace. Further, once the guidelines were in place, most field directors did little to advertise the program in their districts. One field representative even went so far as to respond to an inquiry from a community-based organization that the program did not exist. A major part of the staff's hesitancy to implement the program was because its goals were viewed as being under the rubric of social welfare and not economic development (see Fitzgerald, 1991a).

In summary, the overarching theory that guides practice holds that assisting firms in adopting new technology and providing the "human capital" training necessary to do so are appropriate economic development activities. This corresponds to what Marris (1982: 130) refers to as the corporatist paradigm in which "the primary task of government is to make the most of the land, labor, skills, raw materials, infrastructure and social amenities within its jurisdiction." Government's role is to act in an entrepreneurial capacity to ensure the profitability of firms within its jurisdictions. Further, Marris continues (1982, p. 132),

> The paradigm assumes, then, that there is a dominant structure of economic organization to which societies must adapt themselves competitively, that social goals incompatible with competitive success are unrealistic and self-defeating.

As the Ohio example illustrates, this focus has created an artificial separation between economic and social goals in economic development practice.

Although debate over whether the role of schools is to educate well-rounded individuals or to prepare workers for the labor market has been long standing (see Gray, 1988; Grubb, 1987), the same corporatist model dominates much of education practice. Dippo et al. (1991: 6) define a corporatist approach to education as "an approach that emphasized the production, organization and regulation of human capacities to fit the existing social and technical relations and material conditions of the workplace."

This approach is particularly evident in the community colleges. Increasingly, they are expanding their economic development role through business development centers offering customized training for individual companies. There is little uproar among faculty or administrators that this is an inappropriate function. Nor are taxpayers revolting because this function represents public sector assumption of a private firm's responsibility. Yet Illinois Governor Jim Edgar's recent proposal to transfer the administration of Project Chance, a welfare-to-work training program, from the Illinois Department of Public Assistance to the state's community colleges was met with little enthusiasm by the college administrators. While there are valid reasons for their hesitancy, not the least of which is a real lack of resources, a comment from the state's director of public aid reveals that perhaps community colleges are not very interested in training the most disadvantaged:

> If Los Angeles has any message for us, it's that we've got to be serious about moving people from welfare to work. One part of being serious is to say to community colleges, "Doggone it, you were set up to train people, and it's not just middle-class white people who ought to be on your campuses. It's people who are truly in need and want to learn to work who ought to be there too."[6]

The situation parallels that of the HUPP. The issue is not whether the community colleges should be in the business of training, but who is being trained. The education community seems to view efforts to reach those most educationally disadvantaged as a social welfare, and not an education, function. The next section examines the degree to which recent approaches to linking education to the economy are guided by the corporatist model.

Education as Economic Development

A major failing of the U.S. education system is in facilitating school-to-work transitions for students not entering college (William T. Grant Foundation, 1988; Salaman, 1989). In contrast, several European countries offer non-college-bound secondary school students a range of training options to move them into the workplace: apprenticeships, work cooperative programs, and technical training programs (Haveman and Saks, 1985; Hamilton, 1990; Hilton, 1991). These programs provide degrees and "portable" technical skills that may be used in a variety of companies and sectors. For the 80% of U.S. students who do not complete college degrees, there are few programs to facilitate the school-to-work transition. Many of these students will be consigned for life to low-wage, unstable jobs in the secondary labor market.

The inability to provide the academic and applied skills necessary for successful school-to-work transitions is especially evident for those most in need. A recent congressional study on vocational education discovered that public schools with high percentages of low-income students have poorer quality academic and vocational education programs (Wirt, 1991). Further evidence of the low quality of vocational education is that there is no difference in earnings or placement between high school graduates in vocational and general tracks (Osterman, 1987). A recent study of programs in Chicago revealed that fewer than 10% of high school vocational education students acquire enough skills for employment (Orfield, 1990). Throughout the United States, vocational programs have collected those students who were unable to succeed in the college-prep curriculum.

Many school reform initiatives focus on technical preparation (tech-prep) for high school students.[7] Tech-prep is a program through which two years of high school study are followed by two years of postsecondary study (community college or apprenticeship; this is referred to as a 2+2 model). In some cases, programs continue at four-year institutions (2+2+2 model; see Hull and Parnell, 1991). Much of the money available for planning and implementing tech-prep programs in the schools comes from funds designated under the federal Carl D. Perkins Vocational and Applied Technology Education Act.

Tech-prep programs being implemented using federal Carl Perkins funds are more comprehensive than existing vocational programs. Rather than focusing on narrowly defined job skills, students learn about all aspects of an industry, including planning, management, financing, technical and production skills, underlying principles and technology, labor and community issues, and health, safety, and environmental issues (Wirt, 1991; Rosenstock, 1991). To achieve competency in these areas, students also must be proficient in the basic components of the academic curriculum. This merging of academic and vocational education ensures that students have the necessary background to pursue a college degree, if they so choose, after graduation. To ensure that the curricula developed are responsive to the needs of the workplace and the community, the legislation mandates that planning and implementation be undertaken by partnerships comprising educators and business, labor, and community representatives.

The stated goal of the Carl Perkins legislation, "increasing the economic position of the United States in the world economy by upgrading the skills of the labor force," reveals its orientation in the corporatist paradigm. Yet, at the same time, a key focus of the legislation is on ensuring access to vocational education programs for members of "special populations." To achieve this goal, funds are allocated for minority and sex equitable access and there are programs for single parents, displaced homemakers, single pregnant women, and criminal offenders. In fact, funds are targeted to sites or programs that serve the highest concentrations of individuals who are members of special populations.

Yet the reality of implementation promises to be dismal in many places. There are two problems. The first is the degree to which teachers and administrators "buy into" the reform. Their apprehension toward the policy is very different than that of the economic development planners implementing the HUPP program but has the same potential in blocking its implementation. The second is the lack of resources available for effective implementation.

The second-class status assigned to vocational students frequently holds for vocational education teachers. Because it calls for such a sweeping overhaul of the education system, requiring teachers to work in partnership with business and labor representatives, "second-rate" teachers beleaguered by the system feel threatened by the new demands being placed upon them. Indeed, the legislation is viewed not as a way to improve educational quality but as a top-down Washington initiative to create more work. For example, a recent letter from the editor of *Vocational Training News* warns potential subscribers that "Washington has some nasty surprises in store for Voc Ed administrators" and continues that "New [Carl D. Perkins] regulations require costly evaluations for school districts and their voc ed programs. You'll have to meet stricter performance standards than ever before." Further, most teachers have never shared curriculum development responsibilities and many do not welcome the turf invasion by business, labor, parents, or local residents.

The Carl Perkins Act has little potential for improving education for "special populations" unless it is accompanied by fundamental reform in the way education is funded. Though Carl Perkins funds are targeted to help "special populations," the amount of resources expended cannot compensate for huge disparities between school districts between and within states. In Illinois, for example, local property taxes pay for 62.8% of per pupil eduction expenditures; state funds, for 31.5%; and 5.7% comes from federal programs. Both state and federal funds are used to equalize disparities in per pupil expenditures caused by variations in land value and tax rates. But the low proportion of state money going into education ensures that equality cannot exist. In Illinois, funds allocated to distributive aid were reduced in 1978 and 1982. The result was to increase support for districts with moderate levels of disadvantaged students and to decrease aid to districts with high concentrations of low-income students (Hess et al., 1991: 3). Money flowing into low-income areas is declining while the percentage of low-income students is growing.

Targeted federal legislation, such as the Carl Perkins Act, does not make up the difference. A typical Chicago high school to which Perkins tech-prep funds would be targeted has a 25% graduation rate, has a high truancy rate, is 90% black, and draws 85% of its students from low-income households. With a $50,000 tech-prep planning grant, this school is expected to create a new curriculum that covers all aspects of an industry and links the vocational program to the academic program in such a way that students graduating will be prepared for any postsecondary option they choose.

While the Perkins legislation funding provisions create the potential for greater equity, careful scrutiny and enforcement will be necessary if they are to be followed.[8] If the new tech-prep initiatives are going to make a difference, radical reform in school financing at the state level has to go along with them. Some states, such as Oregon, are making this commitment, but overall it is not politically popular.[9]

The examples presented illustrate that it is difficult to incorporate equity objectives into the corporatist model. Yet, in fact, as the skills mismatch studies document, the U.S. economy will not be able to remain competitive if large segments of the population are denied access to quality education and training. Two cases are presented below as examples of how education and training reform informed by an equity model of work force education and training might look. Its key elements include local participation in defining existing problems and their solutions and equal access to quality education.

An Alternative Model of Training as Economic Development: The Machine Action Project[10]

Historically, the Springfield, Massachusetts, area economy has been dominated by large metalworking firms. Economic development activity focused on attracting these and other large manufacturing firms. Between 1979 and 1986 the Springfield, Massachusetts,[11] area lost over 9,500 metalworking jobs. An unnoticed trend occurring during the same period was the growth of small metalworking firms requiring highly skilled labor. While workers in the large plants were being laid off, positions in smaller firms were going unfilled because of a lack of qualified applicants. It took a union-led initiative, the Machine Action Project (MAP), to respond to this skills mismatch. Without the MAP initiative, this thriving manufacturing sector easily could have been lost and counted as another example of the "inevitable" decline of manufacturing.

There was little response to plant closings from the economic development community, local government, or labor. It was not until 1986, when the United Technologies (UT) Diesel Systems plant (formerly American Bosch) announced its closing, that a local movement to stop the closings emerged. UT workers noticed many signs of disinvestment prior to UT's closure announcement. The union tried to warn local economic development actors and local, state, and national elected officials, but all believed UT's repeated statements of commitment to stay.

Frustrated by the union's lack of influence over responses to this and other local plant closings, Bob Forrant, the union's business manager, organized a coalition to develop a more proactive response to local job loss. The coalition included other union locals, community organizations, training providers, high schools, and community colleges. Their collective efforts coalesced in a proposal for designation as a Cooperative Regional Industrial Laboratory (CRIL), a program created to retain key industries in the commonwealth. In 1986 funds were granted for the formation of the MAP as a CRIL.

The MAP immediately began a study of job loss in the metalworking industry. It revealed a striking pattern in the job loss that had occurred. All but two of the forty metalworking plant closings were branch plant relocations. Even though only 8% of Hampden County's metalworking firms were large branch plants, they accounted for one-third of metalworking jobs. Further, the study identified over 350 small metalworking shops in the area. In contrast to the highly automated large facilities, these plants were thriving, employing approximately 15,000 workers in plants using modern flexible production techniques. The presence of so many vibrant metalworking firms was surprising. Local economic development activity had focused on the needs of large plants and on industrial attraction efforts—primarily the development of industrial parks.

Awakened to the importance of these small facilities to the area's economy, the MAP began a study of their skill needs. It revealed a previously unidentified skills mismatch. The business of the smaller firms was driven by markets in several high-tech sectors, including defense, aerospace, medical equipment, computers, and machine tools. Serving this wide customer base required periodic changes in machine set up as well as ongoing research in product and process development. Workers in these small facilities needed skills in setting up and operating three or more machine tools, reading and assembling parts from blueprints, and maintaining their own tools.

Thus a major concern of the small producers was finding skilled labor. Few of the workers displaced from the large automated plants had the skills needed in the small facilities. Few new labor force entrants were considering metalworking jobs, due to a local public perception that metalworking was a dying industry. While public perception was one of decline and unstable employment, both job security and advancement potential in fact were better in smaller firms than in larger ones.

Despite their concerns, the small firms were not planning or committing financial resources to training (Cann et al., 1991). Vocational programs in the metalworking trades were available locally through the Massachusetts Career Development Institute and five vocational high schools. The programs were offered only on a full-time, fixed schedule basis and thus were not viable options for employed people seeking skills upgrading.

In 1988 the MAP began working with local employers, schools, and colleges, training providers and community organizations to develop a coordinated program for delivery of education and training in the machine trades through five vocational high schools in the region. The courses developed were targeted to upgrade the skills of those already employed in area metalworking shops. The MAP signed a memorandum of agreement with Springfield Technical Community College, the local chapter of the National Tooling and Machining Association, and several government agencies on a common curriculum to be used by all metalworking training institutions.

The program developed served the needs of the area's small metalworking shops, the employees, and the schools. The firms paid a minimal fee of $100 per employee per class for state-of-the art instruction they did not have the resources to provide on their own. The coordination of courses, a common intake assessment, and a competency-based curriculum allowed trainees to take courses with different providers without losing program continuity. Further, the articulation agreement between the schools allowed trainees to accumulate credits toward an associate's degree in metalworking technology. The state funds and tuition were used to update the schools' curricula with one recently updated by the National Tooling and Machining Association and to buy equipment. Further, the MAP cosponsored a week-long Summer Institute on new manufacturing technologies for local secondary vocational education teachers to keep them updated in the latest manufacturing technologies.

The MAP also developed a training program to increase the presence of women in the machine trades. To facilitate completion, the machining program was redesigned to be self-paced, allowing completion in 20 to 40 weeks. This eliminated rigid schedules that made it difficult for working women or those with families to attend. A workshop, "Choosing a Living Wage," was developed to expose women to career options in nontraditional occupations. Videos and brochures, in both English and Spanish, were developed to further increase awareness of the machining trades. A support group for female trainees was organized to provide a forum for women to discuss their common problems and experiences in working in male-dominated occupations. These efforts resulted in an eightfold increase in women's enrollment in machining training programs.

Realizing that the local economy had to diversify, the MAP began to identify other potential growth industries. Two local industries, printing and graphic arts and auto mechanics, were found to have skills mismatch problems. These mismatches had many of the characteristics of the mismatch in metalworking. In response, the MAP organized Project CREATE (Cooperative Resources to Enhance Access to Jobs Through Technology Education) in 1988. The project is funded by two grants, totaling $750,000, from the Department of Education's Office of Vocational and Adult Education. The programs target displaced workers, minorities, disadvantaged youth, women (particularly displaced homemakers), and new immigrants. A unique aspect of this initiative was that workers provided substantial input into both background research and program development. A task force with representatives from schools, training institutions, and industry and other advisers developed the program curricula. Workers in the two targeted industries were surveyed and shadowed to determine how their jobs had changed, and would change in the future, as a result of new technologies being incorporated into the workplace.

Again the MAP entered into a memorandum of agreement with several government agencies, local schools, and training providers to develop a mutual competency-based curriculum in graphic arts and printing technology. The curriculum consists of both classroom and apprenticeship training. Workers serve on the advisory boards of the education and training providers. Their participation provides workers a voice in course content and teaching methodologies and encourages worker participation in skills upgrading.

Three elements of the MAP's approach collectively form a model of how training can drive local economic development strategies. First, employment training is linked to well-defined economic development goals. Traditionally, economic development and employment training are organized under separate planning or local government departments. Employment training divisions primarily are charged with administering federal JTPA programs. JTPA charges local private industry councils (PICs)—representatives from local businesses—with the responsibility of determining local training needs. While there is a great deal of variation in how active PICs are in their role, in too many cases the PIC simply rubber stamps decisions made by the program administrators. The result is training programs that produce graduates with skills that are not in demand in the local economy or with skills that only qualify workers for low-paying service sector jobs and bear little relation to the skills they already possess (Baumer and Van Horn, 1985; Chisman, 1989).

Rather than the question, "What's growing and how can we get it," the MAP's efforts start with a different question: "How can we build on what we have to get to where we want to be?" This labor force-based approach (see Ranney and Betancur, 1992) builds upon the strengths of the local labor force and uses available training resources as part of a broader strategy of maintaining living-wage employment, particularly in manufacturing. In the MAP's case, the strength of the local labor market was the metalworking industry and its labor force. If the MAP had not intervened, the skills mismatch would not have been identified, and, in all likelihood, the small firms would have begun closing due to lack of skilled labor.

Second, worker involvement in the research and development of training programs has resulted in programs more responsive to the needs of both workers and firms. Organized labor historically has not been a recognized actor in local economic development. Nor are the solutions they propose, such as early warning systems and eminent-domain buyouts, familiar in traditional economic development practice (see Fitzgerald, 1991b; Fitzgerald and Simmons, 1991). In fact, they typically are viewed as "antibusiness." Likewise, workers historically have had little influence in determining how to organize production.

Inclusion of workers netted the MAP insights that would not have been obtained otherwise. For example, survey respondents reported that there was little communication between management and labor on how to incorporate new technologies. One result is underuse of new technology. The MAP survey found that 60% of CNC machines and 40% of CAD equipment in the Springfield area were underused. The survey results demonstrate that adopting advanced technology does not necessarily mean that workers will benefit. In many cases, the highly skilled work was being completed by a small cadre of engineers, managers, and consulting personnel while the majority of frontline workers received no training. Indeed, the Project CREATE report concluded that "many new technologies are in fact, if not in specific intent, deskilling large groups of blue collar workers by shifting highly skilled work to engineers, managers, and outside personnel, thus removing a measure of skill and power from the shop floor" (McGraw and Forrant, 1992: 20).

Third, training was used to increase the access to skilled trades of minorities and women. Unions historically have worked to exclude both groups from their ranks. The MAP's commitment to targeting women and minorities for training in occupations that provide living-wage employment demonstrates a new cooperation between labor and various interests in the community in achieving their mutual goals.

An Alternative Model of Education as Economic Development: The Youth Enterprise Network

The Youth Enterprise Network (YEN) demonstrates how vocational education can be incorporated into a broader community development strategy. It is a vocational program in business ownership operating in three Chicago high schools. The program is a collaborative school reform effort, undertaken by Bethel New Life, Chicago Cities and Schools, and the Chicago Workshop on Economic Development (CWED) with the three high schools and the assistance of the Center for Law and Education, a national nonprofit policy and advocacy organization in vocational curriculum reform.

The three local organizations involved in the YEN were looking for ways to use schools as engines of community development. The three schools identified in the Garfield Park neighborhood had high dropout rates and were not incorporated into any efforts to reduce the unemployment, poverty, and general degeneration that characterized the area. Bethel New Life and the CWED had been working on several community-based development initiatives and linked with the Chicago office of Cities and Schools, a nationally based group seeking to reduce dropout rates in inner-city schools, to form the YEN. The program began by preparing the teachers for developing and teaching the new curriculum. Teachers took seminars in entrepreneurship at Chicago State University and explored innovative approaches to teaching. The curriculum developed provides YEN students with the academic and applied skills needed to own and operate a business. It follows the Carl Perkins Act "all aspects on an industry" guidelines, preparing students for any number of careers within an industry. Each year the necessary writing and mathematics skills are incorporated into the students' academic courses.

Students learn the necessary skills to conduct market research, do feasibility analyses, develop business plans, and manage business operations. Students apply these skills by doing a local market analysis, developing a business plan, and setting up a business within the school. The focus of these activities is on identifying business opportunities within the neighborhood. In the three schools involved, businesses started include a school supply store, desktop catering, a credit union, a secondhand infant clothing store, and a greeting card shop featuring the designs of art students.

The program involves the broader community. Neighborhood residents and parents of students enrolled in the YEN are eligible to take courses in small business development. Local business people are being recruited to serve as mentors and provide students real-world examples of business operations.

The critical difference between the YEN approach and similar programs led by business is that entrepreneurship is advanced not as a ticket out of inner-city neighborhoods but as a means of community development. The organizations developed the program because they had identified schools as one of the few neighborhood institutions with the facilities and human resources needed for community development. Other institutions are being drawn in as well. The Center for Urban Economic Development at the University of Illinois at Chicago has been conducting workshops with the students to familiarize them with the broader community economic development process. Eventually, even broader reforms, such as restructuring the school day to make the school more available to youth, families, and community members, are being planned.

Conclusion

The numerous reports on skills mismatch argue that U.S. firms do not excel at technology adoption and skills upgrading. Yet the favored solutions call for bringing the legion of largely ineffective (at training) businesses into the schools to help shape more labor market responsive education programs. As conceived under the corporatist model, the term *skills mismatch* connotes a business sector hampered in its desire to employ the latest production technologies by a labor force that does not have the necessary skills. The corporatist solution is to restructure education to better respond to the needs of business. Once the private sector can be assured an adequately trained labor force, incorporation of new technologies will follow. It is assumed that labor market inequality and discrimination will be resolved once those currently left out funnel through the new education and training programs.

In fact, the problem is more complex than the model suggests. Technology can be incorporated so as to reduce the skill requirements of workers, and workers have little input into such decisions. Corporations can choose to move out of low-skill, inner-city labor markets, leaving residents with few other employment options. At the same time, corporations are more involved in schools than ever. Somewhere between 40% and 75% of U.S. public schools have some link with the business world (Timpane and McNeill, 1991). It is doubtful, however, that these efforts will increase employment opportunities for those most at risk. For example, Public/Private Ventures, a not-for-profit organization that designs, manages, and evaluates school-business partnerships, recently evaluated nine model school-business partnerships targeting disadvantaged youth and concluded:

> As a group, systemic programs have increased attendance rates and report limited positive effects on academic achievement scores; however, they have been unsuccessful in reducing high dropout rates. Furthermore, some programs produced no measurable effects on academic variables; even when changes were found, it was often impossible to directly credit the partnership program. (Public Private Ventures, 1987: 56)

Even larger systemic reforms of vocational education such as the Carl Perkins Act, while needed, may only perpetuate existing class inequalities and racial exclusion. For example, in a discussion on improving the quality of public schools, one Chicago businessman argued:

> It doesn't make sense to offer something that most of these urban kids will never use. No one expects these ghetto kids to go to college. Most of them are lucky if they're even literate. If we can teach them useful skills, get them to stay in school and graduate, and maybe into jobs, we're giving them the most they can hope for.[12]

One can only conclude from such comments that second-rate vocational programs are considered good enough for inner-city minority students. The corporate model does not call into question the contradiction of businesses "adopting" inner-city schools while moving their operations to the suburbs in search of better educated workers. Addressing skills mismatch through new tech-prep or adopt-a-school programs while ignoring the role of spatial mismatch and discriminatory hiring practices will result in a better educated middle class but still leave inner-city minorities excluded from all but the most menial of occupations. What the corporatist model praises as a "best practice" approach is in fact a white, middle-class, suburban solution.

The point is not that business should not be a partner in shaping education reform. Many firms are making good faith efforts to reach at-risk students. But the corporatist approach does not address the inherently unequal distribution of education and training funds within firms. Firms cannot continue spending three-fourths of their training dollars on college-educated management staff as opposed to frontline workers. States cannot continue distributing meager Carl Perkins Act dollars to schools while allowing five-to-one per pupil funding disparities between local school districts to exist.

The MAP and YEN initiatives together demonstrate an equity model of education and training. They demonstrate that reforming our education and training systems to be more responsive to the needs of the economy does not require them to continue past patterns of discrimination and exclusion. Further, the YEN network demonstrates how the Carl Perkins legislation can be implemented to ensure access and equity.

The MAP and YEN initiatives illustrate that blue-collar workers and inner-city residents, labor and community, can participate in defining their problems and shaping effective solutions. Both coalitions recognize that their solutions cannot continue to be funded by foundation grants and specially targeted funds from federal programs. Thus each works to build local constituencies for broader policy change. The MAP is an affiliate of the Federation for Industrial Retention and Renewal (FIRR), a national organization pursuing local strategies for maintaining and modernizing manufacturing jobs. YEN coalition members are working to support a constitutional amendment in Illinois that would reduce significantly the huge disparities in education spending among local school districts. Only when these disparities are addressed can we create the labor force of the future.

Notes

1. See Davidoff (1965), Clavel (1986), and Krumholz and Forester (1990).

2. For example, much has been published on equity planning under Mayor Harold Washington in Chicago. See Mier et al. (1986) and Clavel and Wiewel (1991). For a broader review of African American mayors and equity planning, see Mier et al. (1992). For a discussion of how industrial decline has prompted demands by labor for inclusion in the planning process, see Fitzgerald (1991b) and Fitzgerald and Simmons (1991).

3. See, for example, Clifton et al. (1989), Atkinson (1991), and Shapira (1990).

4. The links between skills mismatch, spatial mismatch, and discrimination are difficult to untangle. Spatial mismatch is caused by an increasing concentration of high-skill jobs in information-intensive industries in the cities and expansion of manufacturing and lower skill service jobs in the suburbs and exurbs (Kasarda, 1983; Blackley, 1990; Cain and Finnie, 1990; Ihlanfeldt and Sjoquist, 1990; Moore and Laramore, 1990). Yet Goldsmith and Blakely (1991) present evidence showing that neither increased access to transportation to suburban jobs nor migration to areas of suburban job growth will overcome racial imbalance. Indeed, the fact that unemployment rates for blacks are more than twice as high as for whites at all education levels (Salamon, 1989) suggests that spatial and skills mismatch are interconnected with racial discrimination (Farley, 1987; Lichter, 1989; Goldsmith and Blakely, 1992).

5. Other states have developed similar programs. The Minnesota Employment and Economic Development (MEED) program provides wage subsidies to firms that hire the

long-term unemployed into on-the-job training positions. Penalties are applied if the firm does not hire the individual after the training period.

6. This is quoted from *The Chicago Tribune* (Dellios, 1992: 17).

7. Apprenticeships are being expanded, but much less so than tech-prep. In apprenticeships, students spend part of their week in the classroom and part learning on the job. The curriculum is defined by employers, and students receive wages while working. Most of apprenticeship programs available in the United States are operated by unions, and not the schools.

8. Lauren Jacobs, staff attorney at the Center for Law and Education, points out that funding mandates of the Perkins Act address intradistrict equity. Funds can only be used in programs that provide equitable participation for special populations, consistent with specific requirements for (a) providing access to the full range of vocational programs available to other students, (b) not discriminating on the basis of special population status, and (c) providing the support services needed by those students to succeed in the program. Department of Education regulations try to undermine these requirements and are being challenged in the courts by the Center for Law and Education.

9. A voter initiative, Measure 5, was passed in Oregon to require state government to take primary responsibility for funding local schools. State assumption of education will take place in five steps. A similar proposal, calling for the state to assume half of local school financing, will be voted on in Illinois in November. It is likely that such proposals will be more difficult to pass in states such as Illinois, where there are larger suburban-urban disparities in school funding.

10. This section draws upon Fitzgerald and McGregor (1993).

11. Springfield, with a population of 150,000, is the center of the five-county Pioneer Valley, with a total population of approximately 600,000.

12. Cited in Kozol (1991: 76).

References

Atkinson, Robert D. (1991). "Some States Take the Lead: Explaining the Formation of State Technology Policies." *Economic Development Quarterly* 5: 33-44.

Bailey, Thomas (1991). "Jobs of the Future and the Education They Will Require: Evidence from Occupational Forecasts." *Educational Researcher* 20: 11-20.

Baumer, D. C. and C. E. Van Horn (1985). *The Politics of Unemployment.* Washington, DC: Congressional Quarterly Press.

Bell, Daniel (1973). *The Coming of Post-Industrial Society.* New York: Basic Books.

Blackley, Paul R. (1990). "Spatial Mismatch in Urban Labor Markets: Evidence from Large U.S. Metropolitan Areas." *Social Science Quarterly* 71: 39-52.

Braverman, Harry (1974). *Labor and Monopoly Capital: The Degradation of Work in the Twentieth Century.* New York: Monthly Review Press.

Cain, Glen G. and Ross E. Finnie (1990). "The Black-White Difference in Youth Employment: Evidence for Demand-Side Factors." *Journal of Labor Economics* 8: S364-S395.

Cann, Elyse, Kathleen McGraw, and Robert Forrant (1991). *Phoenix or Dinosaur: Industrial District or Industrial Decline.* Springfield, MA: Machine Action Project.

Chisman, Forrest (1989). "An Effective Employment Policy: The Missing Middle" in Lee D. Bawden and Felicity Skidmore (eds.), *Rethinking Employment Policy* (pp. 249-264). Washington, DC: Urban Institute Press.

Choate, Pat and J. K. Unger (1986). *The High Flex Society.* New York: Knopf.

Clavel, Pierre (1986). *The Progressive City.* New Brunswick, NJ: Rutgers University Press.

Clavel, Pierre and Wim Wiewel (1991). *Harold Washington and the Neighborhoods.* New Brunswick, NJ: Rutgers University Press.

Clifton, David S., Jr., Larry R. Edens, Harris T. Johnson, and Robert W. Springfield (1989). "Elements of an Effective Technology Assistance Policy to Stimulate Economic Development." *Economic Development Quarterly* 3: 52-57.

Cotton, J. (1985). "More on the 'Cost' of Being a Black or Mexican American Male Worker." *Social Science Quarterly* 66: 867-885.

Cyert, Richard M. and David C. Mowery (1987). *Technology and Employment Innovation and Growth in the U.S. Economy.* Washington, DC: National Academy Press.

Davidoff, Paul (1965). "Advocacy and Pluralism in Planning." *Journal of the American Institute of Planners* 31: 186-197.

Dellios, Hugh (1992). "Foes Say Edgar Dimmed Aid Recipients' Hope." *The Chicago Tribune* (June 13): 17.

Dertouzos, M. L., R. K. Lester, and R. M. Solow (1989). *Made in America: Regaining the Productive Edge.* Cambridge: MIT Press.

Dippo, Don, Arleen Schenke, and Roger I. Simon (1991). *Learning Work: A Critical Pedagogy of Work Education.* Westport, CT: Bergin and Garvey.

Farley, J. E. (1987). "Disproportionate Black and Hispanic Unemployment in U.S. Metropolitan Areas: The Roles of Racial Inequality, Segregation and Discrimination in Male Joblessness." *American Journal of Economic Sociology* 46: 129-150.

Fitzgerald, Joan (1991a). The Effectiveness of Ohio's JTPA System. Report submitted to the Ohio Bureau of Employment Services and JTP Ohio, Columbus.

Fitzgerald, Joan (1991b). "Class as Community: The New Dynamics of Social Change." *Environment and Planning D: Society and Space* 9: 117-128.

Fitzgerald, Joan and Allan McGregor (1993). "Labor-Community Initiatives in Worker Training in the United States and the United Kingdom." *Economic Development Quarterly* 7:172-182.

Fitzgerald, Joan and Louise Simmons (1991). "From Consumption to Production: Labor Participation in Grassroots Movements in Pittsburgh and Hartford." *Urban Affairs Quarterly* 26: 512-531.

Form, William, Robert L. Kaufman, Toby L. Parcel, and Michael Wallace (1988). "The Impact of Technology on Work Organization and Work Outcomes" in George Farkas and Paula England (eds.), *Industries, Firms, and Jobs* (pp. 303-328). New York: Plenum.

Goldsmith, William W. and Edward J. Blakely (1991). *Separate Societies.* Philadelphia: Temple University Press.

Gray, Kenneth C. (1988). "Vocationalism Revisited: The Role of Business and Industry in the Transformation of the Schools." *Journal of Vocational Education Research* 13 (4): 437-445.

Grubb, Norton W. (1987). "Responding to the Constancy of Change: New Technologies and Future Demands on U.S. Education" in Gerald Burke and Russell W. Rumberger (eds.), *The Future Impact of Technology on Work and Education* (pp. 118-140). New York: Falmer.

Hamilton, Stephen F. (1990). *Apprenticeship for Adulthood.* New York: Macmillan.

Haveman, Robert H. and Daniel H. Saks (1985). "Transatlantic Lessons for Employment and Training Policy." *Industrial Relations* 24 (1): 20-36.

Hayes, Robert H. and Ramchandran Jaikumar (1988). "Manufacturing's Crisis: New Technologies, Obsolete Organizations." *Harvard Business Review* (September-October): 77-85.

Hess, Alfred G., Jr., Richard D. Laine, and James Lewis (1991). *The Inequity in Illinois School Finance.* Chicago: Ed Equity Coalition, Chicago Panel on Public School Policy and Finance.

Hill, E. W. and H. M. Rock (1989). "Education as an Economic Development Resource." *Environment and Planning C: Government and Policy* 8: 53-68.

Hilton, Margaret (1991). "Shared Training: Learning from Germany." *Monthly Labor Review* 114: 33-37.

Hull, Dan and Dale Parnell (1991). *Tech Prep Associate Degree: A Win/Win Experience.* Waco, TX: Center for Occupational Research and Development.

Ihlanfeldt, Keith R. and David L. Sjoquist (1990). "Job Accessibility and Racial Differences in Youth Employment Rates." *American Economic Review* 80: 267-276.

Johnston, W. B. and A. E. Packer (1987). *Workforce 2000: Work and Workers for the 21st Century.* Indianapolis, IN: The Hudson Institute.

Kasarda, J. D. (1983). "Entry-Level Jobs, Mobility, and Urban Minority Unemployment." *Urban Affairs Quarterly* 19: 21-40.

Kelley, Maryellen R. (1990). "New Process Technology, Job Design and Work Organization: A Contingency Model." *American Sociological Review* 55: 191-208.

Kozol, Jonathan (1991). *Savage Inequalities.* New York: Harper Perennial.

Krumholz, Norman and John Forester (1990). *Making Equity Planning Work: Leadership in the Public Sector.* Philadelphia: Temple University Press.

Lichter, D. T. (1988). "Racial Differences in Underemployment in American Cities." *American Journal of Sociology* 93: 771-792.

Marris, Peter (1982). *Meaning and Action: Community Planning and Conceptions of Change.* London: Routledge & Kegan Paul.

McGraw, Kathleen and Robert Forrant (1992). *Skills, Training and Education in the Automotive Repair, Printing and Metalworking Trades:* Springfield, MA: Machine Action Project.

Mier, Robert, Joan Fitzgerald, and Lewis Randolph (1992). "African-American Elected Officials and the Future of Progressive Political Movements" in David Fasenfast (ed.), *Community Economic Development: Policy Formation in the U.S. and U.K.* (pp. 90-108). New York: Macmillan.

Mier, Robert, Kari Moe, and Irene Sherr (1986). "Strategic Planning and the Pursuit of Reform, Economic Development and Equity." *Journal of the American Planning Association* 52: 299-309.

Milkman, Ruth and Cydney Pullman (1991). "Technological Change in an Auto Assembly Plant." *Work and Occupations* 18: 123-147.

Mishel, Lawrence and Guy A. Teixeira (1991). *The Myth of the Coming Labor Shortage.* Washington, DC: Economic Policy Institute.

Moore, Thomas S. and Aaron Laramore (1990). "Industrial Change and Urban Joblessness: An Assessment of the Mismatch Hypothesis." *Urban Affairs Quarterly* 25: 640-658.

National Center on Education and the Economy (1990). *America's Choice: High Skill or Low Wages.* Rochester, NY: Author.

Office of Technology Assessment (1990). *Worker Training: Competing in the New International Economy.* Washington, DC: U.S. Congress.

Orfield, Gary (1990). "Wasted Talent, Threatened Future: Metropolitan Chicago's Human Capital and Illinois Public Policy" in Joseph B. Lawrence (ed.), *Creating Jobs, Creating Workers* (pp. 129-160). Chicago: University of Illinois Press.

Osterman, Paul (1987). *The Peace of Vocational Education in a National Employment Policy.* Cambridge, MA: Sloan School of Management.

Public Private Ventures (1987). *Allies in Education: Schools and Businesses Working Together for At-Risk Youth.* Philadelphia, PA: Public Private Ventures.

Ranney, David C. and John J. Betancur (1992). "Labor Force Based Development: A Community Oriented Approach to Targeting Job Training and Industrial Development." *Economic Development Quarterly* 6: 286-296.

Roberts, Markley (1984). "A Labor Perspective on Technological Change" in George W. Taylor and Frank Pierson (eds.), *New Concepts in Wage Determination* (pp. 299-327). New York: McGraw-Hill.

Rogers, Joel and Wolfgang Streeck (1991). *Skill Needs and Training Strategies in the Wisconsin Metalworking Industry.* Madison: Center on Wisconsin Strategy, University of Wisconsin–Madison.

Rosenstock, Larry (1991). "The Walls Come Down: The Overdue Reunification of Vocational and Academic Education." *Phi Delta Kappan"* (February): 434-436.

Rumberger, Russell W. (1987). "The Potential Impact of Technology on the Skill Requirements of Future Jobs" in Gerald Burke and Russell W. Rumberger (eds.), *The Future Impact of Technology on Work and Education.* New York: Falmer.

Salamon, Lester M. (1989). "Overview: Why Human Capital? Why Now?" in David M. Hornbeck and Lester M. Salamon (eds.), *Human Capital and America's Future* (pp. 1-42). Baltimore: Johns Hopkins University Press.

Shaiken, Harley, Stephen Herzenberg, and Sarah Kuhn (1986). "The Work Process Under More Flexible Production." *Industrial Relations* 25: 167-183.

Shapira, Philip (1990). *Modernizing Manufacturing.* Washington, DC: Economic Policy Institute.

Spenner, K. (1985). "Upgrading and Downgrading of Occupations." *Review of Educational Research* 55: 125-154.

Timpane, P. Michael and Laurie Miller McNeill (1991). *Business Impact on Education and Child Development Reform.* New York: Committee for Economic Development.

Wallace, Michael and Arne L. Kalleberg (1982). "Industrial Transformation and the Decline of Craft: The Decomposition of the Printing Industry." *American Sociological Review* 47: 307-324.

William T. Grant Foundation Commission on Work, Family and Citizenship (1988). *The Forgotten Half: Pathways to Success for America's Youth and Young Families.* Washington, DC: Author.

Wirt, John G. (1991). "A New Federal Law on Vocational Education: Will Reform Follow?" *Phi Delta Kappan* (February): 446-455.

Suggested Readings

Baehler, Karen (1987). *Lifelong Learning Part II: Training for a Competitive Work Force.* Washington, DC: Roosevelt Center for American Policy Studies.

Bailey, Thomas (1991). "Jobs of the Future and the Education They Will Require: Evidence from Occupational Forecasts." *Educational Researcher* 20: 11-20.

Beck, E. M. and Glenna S. Colclough (1988). "Schooling and Capitalism: The Effect of Urban Economic Structure on the Value of Education" in George Farkas and Paula England (eds.), *Industries, Firms, and Jobs* (pp. 113-140). New York: Plenum.

Dertouzos, M. L., R. K. Lester, and R. M. Solow (1989). *Made in America: Regaining the Productive Edge.* Cambridge: MIT Press.

Form, William, Robert L. Kaufman, Toby L. Parcel, and Michael Wallace (1988). "The Impact of Technology on Work Organization and Work Outcomes" in George Farkas and Paula England (eds.), *Industries, Firms, and Jobs* (pp. 303-328). New York: Plenum.

Gallo, Frank and Sar A. Levitan (1988). *A Second Chance: Training for Jobs.* Kalamazoo, MI: W. E. Upjohn Institute for Employment Research.

Gold, Lawrence N. (1991). *States and Communities on the Move: Policy Initiatives to Build a World-Class Workforce.* Washington, DC: William T. Grant Foundation Commission on Work, Family and Citizenship.

Hamilton, Stephen F. (1987). "Apprenticeship as a Transition to Adulthood in West Germany." *American Journal of Education* 95 (2): 314-345.

Hamilton, Stephen F. (1990). *Apprenticeship for Adulthood.* New York: Macmillan.

Hornbeck, David W. and Lester M. Salamon (1991). "Overview: Why Human Capital? Why Now?" in David M. Hornbeck and Lester M. Salamon (eds.), *Human Capital and America's Future* (pp. 1-42). Baltimore: Johns Hopkins University Press.

Hull, Dan and Dale Parnell (1991). *Tech Prep Associate Degree: A Win/Win Experience.* Waco, TX: Center for Occupational Research and Development.

Jaikumar, Ramchandran (1986). "Postindustrial Manufacturing." *Harvard Business Review* (November-December): 69-76.

Johnston, William B. and Arnold H. Packer (1987). *Workforce 2000.* Indianapolis: Hudson Institute.

Kazis, Richard (1989). *Education and Training in the United States: Developing the Human Resources We Need for Technological Advance and Competitiveness* (Working Paper). Boston: MIT Commission on Industrial Productivity.

Kozol, Jonathan (1991). *Savage Inequalities.* New York: Harper Perennial.

Langan, Maria (1991). *The Revolving Door: City Colleges of Chicago, 1980-1989.* Chicago: Metropolitan Opportunity Project, University of Chicago.

McMullan, Bernard J. and Phyllis Snyder (1987). *Allies in Education Schools and Business Working Together for At-Risk Youth.* (Available from Public/Private Ventures, 399 Market Street, Philadelphia, PA 19106.)

Mishel, Lawrence and David M. Frankel (1991). *The State of Working America.* Armonk, NY: M. E. Sharpe.

Mishel, Lawrence and Ruy A. Teixeira (1991). *The Myth of the Coming Labor Shortage.* Washington, DC: Economic Policy Institute.

Osborne, David (1988). *Laboratories of Democracy.* Boston: Harvard Business School Press.

Reich, Robert (1987). *Tales of a New America.* New York: Times Books.

Rosow, Jerome M. and Robert Zager (1988). *Training: The Competitive Edge.* San Francisco: Jossey-Bass.

Slessarev, Helene (1988). *Racial Inequalities in Metropolitan Chicago Job Training Programs.* Chicago: Chicago Urban League.

William T. Grant Foundation Commission on Work, Family and Citizenship (1988). *The Forgotten Half: Pathways to Success for America's Youth and Young Families.* Washington, DC: Author.

Wirt, John G. (1991). "A New Federal Law on Vocational Education: Will Reform Follow?" *Phi Delta Kappan* (February): 446-455.

7

Theory and Practice in High-Tech Economic Development

HARVEY A. GOLDSTEIN
MICHAEL I. LUGER

In economic development, as in many other areas of human activity, theory and practice exist as two seemingly separate realities. Academics strive to develop or refine theory, or the pursuit of disciplinary truths (Moore, 1980), by drawing on abstract concepts about the way people behave and institutions work, deriving testable hypotheses about behavior, and testing the hypotheses with empirical data. Practitioners draw from a stock of programs that are "in currency." The theorists do not often specify how their constructs translate into programs, and practitioners rarely ask about the theoretical bases for the program(s) they implement or even the results of empirical testing and evaluation.

Yet, theory and practice are related in a loose, but important way. Theories suggest program "types" (if not specific programs) that are consistent with their beliefs about the way the world works. And, as the world changes, perhaps in small part due to the use of programs, new theories are created and old ones are revised.

The purpose of this chapter is to illustrate the connections between the theory and practice of high-technology economic development. In the second section of the chapter, we review current theories, with particular emphasis on their assumptions and predictions about how they are likely to affect high-tech economic development. The chapter then provides an overview of the types of high-tech development programs that are in use in the United States and other industrialized countries, and discusses their theoretical underpinnings. The fourth section discusses the high-tech development experience of two regions (the "Bionic Valley" in Utah and

AUTHORS' NOTE: We acknowledge assistance in the preparation of the chapter from Stuart Sweeney, a Ph.D. student in our department, and from the book's editors.

North Carolina's "Research Triangle"), widely regarded as "successful," as a way to illustrate the critical link between theory and policy. The last part of the chapter provides a critique of high-tech development policymaking. We conclude that there is not a deterministic, one-to-one mapping between theory and policy. Rather, the most successful policy efforts are those that are built on theories that make sense for that particular region at that particular time, either by design or by happenstance. The most ill-advised programs tend to be those that are "foolishly consistent" with theory or put in place simply for symbolic reasons or because they have been used elsewhere (the "bandwagon" phenomenon).

The review of theory in the following section is not intended to be exhaustive in either breadth or depth. Rather, it covers the schools of thought most prevalent in the economic development literature and most relevant to high-tech development, including regional growth theory, classical location theory, and comparative advantage and trade theory; stage, long wave, and product cycle theories; theories related to the organization of production; and theories of propulsive, innovative, and creative regions. Similarly, our discussion of high-tech development programs in the fourth section is not necessarily complete. We consider programs to recruit high-tech activity from outside the region, to update the technology of existing facilities in the region, to generate and nurture high-tech start-ups, to accelerate R&D, and to create a favorable business milieu in the region. Readers may quibble with what we have included or with the way we have grouped related theories and programs. The basic points we make, however, are not dependent on the particular lists or groupings.

What Is High-Tech Economic Development?

High-tech economic development is a process of business location, creation, and expansion, with an associated set of outcomes. The particular focus of attention is high-technology organizations, including private, for-profit enterprises as well as not-for-profit research-oriented entities. *High tech* has several meanings in the literature; generally, it is used to describe organizations engaged in technically sophisticated activities that lead to product or process innovations, new inventions, or, more generally, the creation of knowledge.[1] Organizations are classified as high tech on the basis of their volume of R&D spending, the percentage of scientists and engineers in their work force, their rate of innovations and inventions, or some combination (for example, see Armington et al., 1983; Glasmeier, 1990). The outcomes of the process of high-tech economic development that are the subject of attention in the literature typically include the gross amount of induced investment that is undertaken, changes in the region's mix of industries, and changes in the number and type of workers that those industries require; technological improvements and productivity gains in existing establishments; and associated political and social shifts.

Theories Related to
High-Tech Economic Development

This section reviews neoclassical economic theories, stage/cycle/wave theories; theories relating to the organization of production; and theories of propulsive, innovative, and creative regions. For each group, we discuss the theories' assumptions and what they predict about a range of economic development outcomes.

Neoclassical Economic Theories

We classify as "neoclassical economic" those theories that focus on market responses to changes or differences in prices. Assuming, as most of the literature does, that markets operate competitively, differences in input prices (for capital, labor, energy, other raw materials, information, and so on) among producers are reflected in differences in output, factor proportions, or both. All factors are conventionally assumed to receive the value of their marginal products as compensation for the services they provide.

Regional growth theory assumes that both capital investment and labor flow freely among regions. The differences in growth rates and wage levels between any two regions result from differences in production technologies. The regions with the most technologically advanced production process will have a higher growth rate. Differences between marginal products of labor and capital in the two regions lead to flows of capital investment into the region where the remuneration to capital is higher and to the migration of labor into the region with higher wages. Over time, the free flow of investment and labor leads to a convergence in regional incomes as marginal products in each region equalize. As long as there are no structural reasons for the original differences in production technology, the equilibrium outcome will be characterized by a set of locations that are identical in terms of industry and skill mix, factor proportions, and other outcomes.

In many cases, production technologies differ for structural reasons. The *theory of comparative advantage and trade*, for example, recognizes that regions have different natural endowments and policy-created strengths (Ohlin, 1933). The basic argument is that a region will specialize in production of the good(s) that use those endowments and strengths. For example, a region rich in fossil fuels will develop an export sector around the petroleum industry. The income it gains from those exports will be used to buy other types of goods produced elsewhere. Society will be better off as a consequence of this trade, as opposed to each area producing a full market basket of goods for its own consumption. Applied to high tech, this means that regions rich in "knowledge resources," such as universities and technical schools and a good information infrastructure, will have a competitive edge in R&D and related activities. One of the things that makes high-tech economic development different is that regions are not naturally endowed with resources but create them through policy over a long period of time.

The theory of comparative advantage and trade assumes the free flow of capital, labor, knowledge, and other factors, among regions. But, because differences among regions cannot be eliminated, at least in the short run, there will not be convergence in factor proportions and industry-skill mix. Instead, factor returns will equalize among regions, but with different industry-skill mixes and factor proportions.

In recent years, *disequilibrium models* based on the ideas of Myrdal (1957) have been developed that stress the process of cumulative causation leading to continued investment in the growing region and a long-term divergence of regional incomes (Malecki, 1991; Malecki and Varaiya, 1986; Vietorisz and Harrison, 1973). Again, capital, labor, and other factors are assumed to flow freely. Because the location of activity depends on the presence of certain key locational features that are continually enhanced by the in-migration of new economic activity, returns to factors may not equalize, in which case the flow of resources from one region to another will not stop.

Growth pole theory, originally formulated by Perroux (1950), states that investments in propulsive industries (the pole) in strategically located centers induce growth by firms in technologically related industries through the formation of backward and forward linkages with the propulsive industries. The "center" is defined as a locus of innovation. In Perroux's original statement of the theory, the induced growth occurred in an "economic region" that did not have to be spatially contiguous to where the propulsive industry was located. In later revisions of the theory, induced growth was within a geographic region and emanated from a growth "center" (Boudeville, 1966; Hansen, 1967: 709-725; Hirschman, 1958; Myrdal, 1957; Darwent, 1969: 5-31).

The types of induced economic growth predicted by growth pole/growth center doctrine are (a) existing firm expansion and new business formation through backward and forward linkages; (b) new business formation in the same industry(ies) as the propulsive industry through localization and agglomeration economies; and (c) firm expansion and new business formation in consumer services and retail trade from indirect and induced growth in residential-based activities. The bulk of the expected induced growth would be through backward and then forward linkages, because the propulsive industries would be chosen in growth pole/growth center doctrine so as to maximize the interindustry trade multiplier effect.

Classical location theory (also called Weberian location theory after the pioneering work of A. Weber, 1929) provides a method for finding the optimal location for a firm, given distances to raw materials and the final market. Producers are assumed to minimize their total costs, including transportation (see Isard, 1956; Beckmann and Thisse, 1986). Markusen et al. (1986) point out that, because high-tech production neither relies heavily on raw materials nor has a high value-to-weight ratio, it does not necessarily use the location calculus specified in the Weberian model.

Another strand in the neoclassical economics literature addresses the process of technological change, particularly *the creation and diffusion of inventions and innovations*. These do not always have an explicit spatial dimension but have clear implications for regional economic development.[2]

The literature on the generation of new technology attempts to determine the role such characteristics as firm size, market power, and R&D activity play in stimulating technological advance among industries. The research also explores the determinants of innovativeness among firms within industries with relatively high rates of technology growth. Particular attention is paid to the patent system and socially optimal methods of compensating inventors (see Taylor and Silberstein, 1973; Scherer, 1977). Arrow (1962) and Dasgupta and Stiglitz (1980), among others, also explore the role of corporate R&D in inventive versus innovative activities.

Research on diffusion addresses the spread of new technology, especially process innovations, within firms (Mansfield, 1968; Stoneman, 1981), between firms (Griliches, 1957; Dixon, 1980; Mansfield, 1968), and, more generally, throughout the economy (Salter, 1966; Chow, 1967; Nabseth and Ray, 1974; Teece, 1977). For each type of diffusion, researchers construct profiles of early and late adopters of the technology, explore the time path of the diffusion (which path and why, and the key factors influencing the path), and consider changes in businesses' use of technology before and after adoption of the technology (see Stoneman, 1983).

Stage/Wave/Cycle Theories

This group includes theories that suggest a recurring, recognizable path or pattern of development over time and/or space. The three major variations focus on stages, long waves, and product cycles. Stage theory focuses on historical, social, political, and economic factors as countries pass through stages of development. Long wave theory was developed mainly in the context of global technological change. The product cycle theory was originally developed in the context of international trade and has an explicit geographic focus. Subsequent to its introduction in the international trade literature, it was embraced and modified by regional development theorists. As its name implies, the theory is concerned primarily with product innovation, although it has been extended to include profit cycles and process innovations.

Stage theory normally is associated with Rostow (1991), who used historical evidence to identify the stages of development and growth that nations experience, from traditional to mature societies with high levels of material consumption. The rate and manner of passage through these stages are affected by economic variables, such as market demand, the propensity to take risks, the level of technology, and the amount of entrepreneurship as well as by political, social, and cultural characteristics. Rostow's empirical analysis suggests that developing countries that have accumulated a critical amount of capital and technology can reach a "takeoff" point in their growth trajectories. Countries' growth and stability beyond that point depend on their ability to harness existing and emerging technologies.

The literature on high-technology regional development does not explicitly cite Rostow. Yet, his insights about the preconditions[3] for takeoff and development into a high-technology-based economy infuse at least two of the literatures reviewed here: on product cycles and regional milieu.

Kondratiev's work in the 1920s is the best known among the early explorations of *long waves*, but the tradition dates back to the work of Parvus in the 1890s. Here, we summarize only the *long wave theories that focus on the importance of innovations and entrepreneurship*. The seminal work on this topic was by Schumpeter (1939), who focused on "irregular clusters" of innovations in the cycles of economic development. Entrepreneurs begin the cycle by introducing several basic innovations into the economy. These basic innovations are things such as iron smelting in the 1780s, the steel making process and the steam engine in the 1840s, and the internal combustion engine in the 1890s. For a short time, they reap large profits by occupying a monopoly position in the market. Development and growth slow as imitators enter the market and slowly compete away the profits. Innovations at this stage amount to improvements of the original innovation. As demand for the innovations becomes saturated and profits disappear altogether, the economy turns downward, into depression, or what Schumpeter called "a period of 'creative destruction,' " ripe for the talents of a new set of entrepreneurs to innovate and start another cycle.

Mensch (1979) provided empirical support for Schumpeter's thesis by statistically associating clusters of basic innovations with the development of new growth industries and the starting points of long periods of development. He used his findings to suggest that the only way out of a depression was through massive government funding to encourage the development of new innovations. Freeman et al. (1982) also saw innovation as a strong force behind the waves of development, but they departed from Mensch by including intraindustry product and process linkages and technology systems as possible generators of development waves.

Vernon (1966) wrote the first comprehensive statement of *product cycle theory* in the context of international trade. He argued that products pass through three phases—new, mature, and standardized—each associated with different markets and production processes, with consequences for the location and types of production facilities in the industry.

External economies are important for the production location decision in the new product stage. So are communications; design of new products is facilitated through constant interaction between the market and producer. A producer needs to see how the market reacts to different design specifications. Given this desire to minimize uncertainty, a producer is likely to locate the first production facilities in the United States rather than in a lower cost, less developed country.

As the product matures, demand for it tends to increase, in part because the designers have found the right market niche through trial and error during the new product stage. At the same time, consumers are able to identify substitutes for the product, so demand becomes more price elastic than it was in the start-up phase. Because the product is no longer unique and flexibility in production is no longer required, producers adopt mass production techniques as a way to exploit scale economies and compete on the basis of price.

These changes in cost structure and the production process have locational implications for high-tech as well as traditional industries. In the new product stage, most demand is from local markets, though there is some exportation to areas with either high income or labor shortages. As the product matures and becomes cheaper and more standardized, demand for it continues to grow in a broader range of external markets. As the level of exports continues to rise, producers begin to move some production facilities to the external markets to reduce costs and to prevent foreign entrepreneurs from setting up their own production facilities and seizing their foreign market share. As the product moves closer to complete standardization, trade dynamics continue to change. Economies of scale for producing in the home market become fully exploited and labor costs increase significantly. More and more production is done in the foreign markets, which have an increasing cost advantage over home production, and, soon, foreign-based subsidiaries begin to export to the home market. In a two-region model, the product cycle is complete when production facilities in the external location import their lower priced goods to the original market and force the closure of the original production facilities.

Markusen (1987) created a variation on product cycle theory by changing some of the original assumptions and focusing on profit cycles. The changes in the original model are as follows: Profit is the cyclical indicator rather than output and demand; imperfect competition (oligopoly) occurs in the maturation stage in addition to the new product stage; the units of analysis are more disaggregated. The dominant spatial outcome predicted by the model is a spatial division of labor. The central contributions of Markusen's work are the alternative theoretical framework and the detail of the analysis and results.

Production Organization

Some contemporary economic geographers draw on Marxist theory, with its focus on the relationship between the form of production and social outcomes, to construct an alternative to the other, more economistic theories reviewed here (for example, see Harvey, 1982; Massey, 1984). Particular attention is given to the division of labor in production, the structure of interestablishment transactional activity, and agglomeration economies due to local forms of development.

Scott and Storper (1987) use this framework to explain the development of "new technology-based growth centers." These develop as an alternative means of capturing economies of scale—through vertical *disintegration* into production complexes rather than through vertical *integration*, which is internal to the firm.

The tendency to form a complex is largely a result of several distance-sensitive transaction costs. For example, large labor markets induce agglomeration because they reduce search costs, offer a larger pool of job seekers, and provide a built-in socialization process. In the case of high-tech complexes, one advantage of this agglomeration tendency is the creation of an environment rich in innovative and entrepreneurial talent and opportunities. At some point in time, however, agglomeration creates diseconomies as the social and political forces that have been created become rigid, conflict with the employers' interests, and increase the cost of production in the region. Technological and organizational changes are required to overcome these diseconomies (Scott and Storper, 1987).

According to this construct, each industrial epoch—including today's high-tech economy—is characterized by an ensemble of productive forces and has a corresponding geography of location. At the dawn of a new epoch, emerging industries will favor locations that have not been the focal point of a previous industrial epoch, where the employment relation can be re-created without the rigidities of the entrenched social-political forces and labor market. Other important features in fostering the growth of new complexes include the local government and infrastructure.

Clark et al.'s (1986) work on the role disequilibrium and uncertainty play in industrial economies provides an alternative to the neoclassical model of regional production, which makes extreme assumptions about space, equilibrium tendencies, and information. Their theory explains how businesses in different sectors of the economy manage the uncertainty they face through short-term adjustments by shifting the instability in the economy onto the least skilled and least organized workers. Corporations search continuously for new pools of labor with no history of militancy or unionization. Outcomes are different for workers in different sectors. The most vulnerable workers in the short run are employed in the tertiary sector (for example, services), which is increasingly automated and has relatively little unionization. Primary and secondary sector workers (especially those in product research and development, early production, and control or management functions) have greater job security in the short run, in part because their production skills have not yet been automated, but greater insecurity in the long run, because of the rapid pace of industrial restructuring.

A related literature, referred to as *post-Fordist*, focuses on the generation of new products and the related innovations in manufacturing processes (see, for example, Aglietta, 1976; Moulaert and Swyngedouw, 1991; Sabel, 1982). The name *post-Fordist* refers to a period that eclipses the era of standardized, assembly-line mass production. Increasingly, for example, products are manufactured in smaller lots and require more R&D as they are customized to meet today's specialized markets. Because these products are relatively costly and have limited shelf life (before becoming obsolete), their manufacturers seek to minimize inventories and maximize the flexibility of the production process. This leads to such practices as just-in-time (JIT) production and component sourcing, which have important spatial implications. For instance, both lead to agglomeration, as component and other intermediate good manufacturers locate near the main production facility. This locational phenomenon lies behind the development of such economic development tools as air cargo complexes and free trade zones.

Theories of Propulsive, Innovative, and Creative Regions[4]

These theories focus on the factors that make some regions dynamic and "synergistic." They place particular importance on the innovativeness and creativity of human agents in the development process. The agents are typically entrepreneurs but also include powerful or insightful politicians, business leaders, university officials, or others. The common question motivating this work is how development can be promoted from within a region. To facilitate discussion, we divide these theories into two groups: those that stress diffusion of growth or innovation, for example, from a center outward, and those that focus on amenities of location that enhance a region's "creativity" or dynamism, without an explicit spatial transmission process.[5] The first group includes growth pole/growth center (this theory was discussed earlier, in the first section under disequilibrium models) and innovation diffusion theories in which relationships among businesses or individuals are of primary importance. These explanations are based on an initial imbalance within the region-at-large. Policies consistent with these theories normally target key (propulsive) industries or innovative individuals. The second group of regional development theories includes entrepreneurship, seedbed, and regional creativity theories in which environmental factors and nonspatial synergies contribute to the area's overall "fertility." These explanations do not necessarily begin with intraregional or sectoral imbalances. Policies consistent with these theories focus generally on individuals or places rather than industries (Pred, 1976: 151-171).

Hierarchical diffusion theory stresses the spatial filtering or trickling down of innovations. But, while the revisions of growth pole/growth center theory focus on the transmission of growth from center to hinterland within a region, hierarchical diffusion theory specifies a "downward filtering through an urban hierarchy" (Berry, 1972, 1973; Lasuén, 1971, 1973: 163-188). Specifically, innovations are transmitted from larger to smaller metropolitan areas. Hierarchical diffusion theory also differs from growth pole/growth center theory in its emphasis on process innovations (Thwaites, Edwards, and Gibbs, 1982; Thwaites, Oakey, and Nash, 1982: 1073-1086; Rees et al., 1985).

Some diffusion theorists have argued that empirical data on interindustry transactions do not support growth pole/growth center or hierarchical diffusion theories of innovation transmission (Pred, 1976; Gilmour, 1974: 335-362; Britton, 1974: 363-390; Erickson, 1975: 17-26; Moseley, 1975). Innovations do not diffuse in a spatially systematic way; rather, intraorganizational linkages within multiplant firms account for many of the transactions of goods and information that are observed. These linkages are typically interregional. Similarly, Higgins has argued that growth pole/growth center and hierarchical diffusion theories are no longer relevant in today's information-age economy. "Today's communications make it perfectly possible," he says, "for an innovation made in Stuttgart to reach Detroit before it reaches Bonn" (Higgins, 1983, p. 7).

An issue in the high-tech development literature relates to exogenous versus indigenous development strategies. (We focus on these strategies in our case study, in the fourth part of this chapter.) *Exogenous development* refers to that which is initiated, propelled, and controlled by organizations located outside the region. *Indigenous development* refers to development that is regionally initiated and planned. It places emphasis on small and medium-sized and regionally owned enterprises rather than on recruited branch plants of large multilocation corporations. The provision of information transfer and advisory services for local propulsive businesses, as opposed to financial incentives for nonlocal firms, for example, would be another attribute of an indigenous strategy.[6] In principle, there is no reason that growth center strategy could not follow an indigenous development path so long as the target region has the requisite resources to start with. There is an empirical and practical question of whether initial investments in an indigenous development strategy would be sufficiently large to generate threshold levels of demand to create forward linkages and agglomeration economies.

The indigenous/exogenous distinction is important for another class of theories, grouped here on the basis of their common focus on the "fertility" of the region as a propagator of new economic activity. The intent of the policies based on these theories is to foster self-generated growth.

Entrepreneurship/seedbed/creativity theorists ask: Why have some locations, such as Austin, Route 128, and Silicon Valley, been economically more dynamic than others, and what role can public policy play in creating and fostering those conditions (Cox, 1985: 17-25; Thomas, 1985)? Andersson (1985: 5-20) provides a general answer: "Creativity as a social phenomenon," he says, "primarily developed in regions characterized by high levels of competence, many fields of academic and cultural activity, excellent possibilities for internal and external communications, widely shared perceptions of unsatisfied needs, and synergies among local actors." He and others suggest that these conditions can be influenced by public policy. These conditions are also related to city size, because larger cities typically provide a wider variety of services and greater agglomeration economies than small cities do.

Modern applications of the net entrepreneurship/seedbed/creativity theory relate the density of entrepreneurs and extensiveness of entrepreneurial networks to rates of growth (Bollinger et al., 1983: 1-14; U.S. Congress, 1985). In particular, these applications stress an "entrepreneurial phase" of development that follows an "institutional phase" (Cox, 1985: 20). During the institutional phase, the region attracts major research facilities, adds services and support operations, and develops a critical mass of scientists and innovators, who begin to interact. Then, during the entrepreneurial

phase, scientists and engineers, singly or in teams, spin off new enterprises, usually located in the same region (Cooper and Komives, 1972:
108-125; Harris, 1986). Policies to increase the density of entrepreneurs
focus either on the potential innovators by providing technical assistance,
peer support, specialized training, or start-up capital, for example, or on
the overall cultural and economic environment (Shapero, 1981: 19-23;
Cox, 1985: 20).

For proponents of entrepreneurship theory, two questions remain out-
standing: whether a sufficient pool of local entrepreneurs exists or can be
developed and whether local economic conditions can support new start-
ups. In some locations, public intervention may be necessary to overcome
the inevitable start-up problems that new firms experience at enormous
costs. In these places, it may be better to attract branch plants of established firms, at least in an initial stage during which essential infrastructure can be developed.

Another critical question for policymakers to ask is whether successful
start-ups will generate sufficient local multipliers in the long run to justify
local support. If they are linked primarily with firms outside the region,
there will be considerable leakage. Then, as we discussed above, higher
levels of government are more appropriate sponsors of policy.

The Relationship Between Theory and High-Tech Development Programs

In practice, state and local high-tech development programs tend to be
ad hoc rather than based upon a particular theory (or combination of
theories) of how regional development is supposed to occur. Those policymakers who are grounded in theory tend not to agree on which explana-
tion(s) should serve as a foundation for action. Time constraints and data
limitations frustrate attempts to test empirically the appropriateness of a
particular theory as a guide for regional high-tech policy. In addition,
because theories are not always well specified, or are presented in highly
abstract terms, policymakers have difficulty linking them to "consistent"
courses of action.

We focus here on this last explanation, attempting to flesh out the
attributes of policies that seem to be consistent with each of the major
theoretical frameworks reviewed above, however incomplete, untested, or
abstruse they may be. We develop a typology of policy approaches based
in practice and identify specific programs that fit within it. We then discuss
the two-way relationship between theory and policy.

High-Tech Development Programs: A Typology

High-tech development programs used by state and local governments can be classified as follows: (a) *recruitment* of high-tech businesses or organizations to locate in the region; (b) *modernization* of process technology in the region's existing production facilities; (c) *incubation and nurturing* of new, "homegrown" high-tech businesses; (d) stimulation of *innovation* in the region, specifically by encouraging R&D; and (e) creation of a favorable *milieu* for entrepreneurial activity and regional synergy.[7] The categories are intended to capture differences in the behaviors of businesses and individual entrepreneurs by which high-tech development actually occurs and how programs are intended to alter those behaviors. It should be noted that, while these categories are meant to be analytically separate, specific programs may fit in more than one category. For example, technology transfer programs can be used to assist existing businesses trying to update their technology or may be targeted to start-up businesses. In the paragraphs that follow, we describe each category of programs more fully.

Recruitment. The major difference between this high-tech policy thrust and the industrial recruitment activities that long have been the backbone of state and local governments' economic development strategies is a change in what is subsidized. To attract net new businesses to a region, recruitment programs must reduce costs that are relatively important to the target businesses. In the case of high-tech industry, those include labor, information, specialized research equipment, and labs. Consequently, high-tech recruitment consists of a smorgasbord of programs that reduce the cost of labor and information, including subsidized job training, employment tax credits, publicly provided telecommunication hookups, and a good transportation infrastructure, and provide government-funded research centers in particular application areas (microelectronics, biotechnology, robotics, and so on). These programs, and other attributes of the state or locality, are publicized to businesses in glossy advertisements and by personal visits by public officials to business executives. North Carolina's Department of Commerce and Tennessee's Technology Foundation have been particularly successful in applying the traditional industry recruitment strategy to the high-tech sector. Some states—for example, Texas—concentrate their recruitment activities on branch facilities of federal government R&D organizations.

Modernization. The economic bases of many states and localities contain a large number of traditional manufacturing businesses that have either closed or shrunk considerably because they have not been able to compete in the global economy. To help that sector, state and local governments have established programs to help existing businesses through the provision of labor training, grants and loans for new equipment, and technical assistance. Well-known examples include Michigan's Modernization Service and Pennsylvania's PENNTAP program. Many states also have developed industrial extension services through their land grant universities.

Incubating and nurturing. These programs have two related thrusts: increasing the number of individual entrepreneurs who start their own high-tech businesses in the region and assisting young, small businesses that may have considerable long-term promise but are vulnerable to failure in their early years due to undercapitalization and limited knowledge about how to run a business. Programs include the provision of venture capital, incubator facilities, and managerial training/assistance. The Massachusetts Technology Development Corporation has been used as a model for other state programs of this type.

Innovation. Incentives are used to encourage businesses and universities to conduct more R&D. These typically include R&D grants and tax credits, building the research capacity of the region's universities, engaging in joint university-industry research/technology centers, and increasing the region's supply of highly skilled scientists and engineers through expansion of university degree programs. The objective is to make the region a center of innovation that will induce the location of more high-tech businesses. Of course, increases in R&D also should facilitate the creation of new processes and products that will enhance the competitive position of the regional and national economies in the world market. The Ben Franklin Partnership in Pennsylvania, the Edison program in Ohio, and the New York Science and Technology Foundation are among the best known state innovation efforts.

Milieu. Creating a milieu that is conducive to creativity, dynamism, and regional synergism can include a wide range of programs that often are not considered in the domain of economic development. The common objective of programs under this umbrella, however, is to attract and retain creative, talented, and risk-taking individuals for the region. Policy/program targets might include the quality and cost of public services, environmental quality, degree of development of organizational networks and cohesion, and quality of community leadership. Many of the regions considered by high-technology business executives to have milieus favorable for entrepreneurial activity also can be considered successful centers of innovation developed around incentive programs (as described above), research parks, favorable locations, or particularly entrepreneurial cultures. The regions often cited for their high-tech milieus include Silicon Valley, Route 128, Austin, and Salt Lake City. Each of these regions contains, or is proximate to, important universities. The development of Silicon Valley can be linked, as well, to the growth of the nearby Stanford Research Park and a particularly pleasant climate and landscape (Krenz, 1987). The development of the Salt Lake City region as a high-tech mecca is due, in part, to a strong entrepreneurial culture (Goldstein, 1991).

We also can classify high-tech programs in terms of their position along four continua: their *demand versus supply* orientation; their emphasis on *capital versus labor*; their *micro versus macro* orientation; and their *exogenous versus indigenous development* focus. *Demand* or *supply orientation* refers to whether the program is meant primarily to affect the supply side of businesses, for example, by lowering factor input costs, or to increase the effective demand for goods produced within the region. The *capital/labor* dichotomy refers to whether the program is targeted to investors or entrepreneurs, for example, by lowering risk, or to labor, for instance, by providing training as a way to increase productivity. A program is micro oriented if it is targeted directly to an individual, enterprise, entrepreneur, or worker, and macro oriented if it is targeted to the local environment to help it become more conducive for high-tech business development. The exogenous/indigenous dichotomy refers to whether programs primarily attract resources that are located, owned, or controlled outside the region (such as equity capital, plant and equipment, and highly skilled labor) versus helping to generate those resources from within the region.

Table 7.1 places specific high-tech development programs that have been used or proposed in practice among the cells of the matrix formed by the two sets of categories discussed above: Program categories are listed on the left side of the table and the four continua are listed at the top.

The distribution of programs among the cells indicates several items of interest. First, high-tech development programs tend to be supply oriented, regardless of policy orientation. That is, the bulk of programs are oriented to reducing factor-input costs to businesses. Second, there is little or no relationship between policy orientation and capital/labor targeting; each of the major policy approaches, in principle, supports program activities that can be targeted to capital or labor. Third, recruitment, modernization, and incubation policy orientations tend to be micro level in approach; they do not differ from their respective non-high-tech economic development relatives other than having an additional focus on technology as an important factor input. Innovation and development of a creative milieu are almost exclusively macro oriented and are distinctive for high-tech development.

TABLE 7.1. Two-Way Classification of High-Tech Development Programs by Policy Approach and Type

Policy Approach	Supply- or Demand-Side Orientation		Capital or Labor Target		Micro- or Macro-Orientation		Exogenous or Indigenous	
	Supply	Demand	Capital	Labor	Micro	Macro	Exogenous	Indigenous
Recruitment of high-tech enterprises into the region	Tax incentives Loans/grants Regulatory relief Investment of public pension funds	—	Tax incentives Access to capital Regulatory relief Loans/grants	Education/training	All programs in row	Infrastructure investments	All programs in row	—
Modernization of process technology in existing regional enterprises	Technology transfer Technical assistance Loans/grants Incubators	Export promotion	Technology transfer/ adoption Technical assistance	Customized training Extension service Technology-based community college programs	All programs in row	Technology transfer Brokers/ organizations Infrastructure investments	—	All programs in row
Incubation and nurturing of technology-based start-up enterprises	Venture capital provision Managerial/ technical assistance	Buyer-supplier networks Export promotion	Venture capital Product development loans Incubators	Customized training	All programs in row	Business services Buyer/supplier networks	—	All programs in row
Innovation by investment in R&D infrastructure	Tax incentives	Technology transfer assistance	Research parks Technology centers Product development loans/grants	Science- and technology-based university degree programs	Grants/research funding to individual R&D organizations		All programs in row	All programs in row
Creation of milieu conducive to regional high-tech development	Venture capital Entrepreneurs' networks	Buyer-supplier networks		Public schools Environmental quality Recreation Affordable housing	—		All programs in row	All programs in row

Fourth, recruiting is, by definition, an exogenous approach, while incubation and, to a large extent, modernization are indigenous approaches. Innovation and milieu creation as policy orientations are neutral with respect to the exogenous/indigenous dichotomy. Fifth, many of the specific programs/activities overlap several of the major policy approaches rather than being within the domain of a single approach. This suggests that it is the *combination* of specific programs, their relative emphases (in terms of support and commitment), and to what entities they are targeted that distinguish major policy approaches among regions and over time.

The Theory-Policy Nexus

Table 7.2 indicates the degree to which each group of programs is consistent with the theories that relate to high-tech development. Here, our judgments are based on the modal cases, recognizing that there is variation among programs within each group as well as from place to place.

To the extent that *neoclassical theory* is consistent with *any* policy intervention—as opposed to suggesting that the "hidden hand" of the market will take care of all problems—it accords with strategies of modernization and R&D investment (innovation) due to its emphasis on investment to increase productivity as a means to gain comparative advantage. *Long wave/stage/product cycle theories* are highly consistent with programs to stimulate regional R&D (innovation) and consistent with incubation and the creation of milieus that will spawn higher rates of entrepreneurial activity. *Theories of production organization* accord most with modernization programs because of their emphasis on changes in both technology and the organization of the workplace. Finally, *theories of entrepreneurship and regional creativity* are highly consistent with programs designed to create favorable milieus and consistent with efforts to stimulate R&D and new business development (incubation). Those theories clearly are inconsistent with recruitment programs.

We offer two tentative conclusions about the relationship between theories and policies of high-tech development from the above discussion. First, there is not a clear-cut one-to-one mapping between high-tech programs and theory. Some policy initiatives are consistent with several theoretical approaches, while others are not consistent with any. Second, a contingent, rather than a general, theory of regional high-tech development is required. A contingent theory would identify the specific theories, with consistent policy approaches, that are most appropriate for a particular region with specific regional conditions and development objectives. We elaborate on this idea below.

TABLE 7.2. Consistency of Theories and Policy Approaches

Major Theories	Recruitment	Policy Approaches			
		Modernization	Incubation	Innovation	Milieu
Neoclassical regional growth interregional trade disequilibrium	0	+	0	+	0
Waves/cycles/ stages	0	0	+	++	+
Production organization	0	++	0	+	0
Entrepreneurship/ creativity	−	+	+	++	0

NOTE: Key: ++ = highly consistent; + = consistent; 0 = neutral; − = inconsistent.

A Tale of Two Regions

A comparison of the cases of Salt Lake City, Utah (sometimes referred to as the Bionic Valley), and the Raleigh-Durham region of North Carolina (now known as the Research Triangle) helps to illustrate how the application of theory to practice is contingent on particular starting conditions and traditions of the particular region.

The Raleigh-Durham region in the late 1950s and the Salt Lake City region in the late 1960s shared several characteristics. Both were moderate-sized metropolitan regions, both contained a state capital, and both had one or more very good research universities. On the other hand, both regions had only mediocre accessibility to major metropolitan centers, a relatively low level of high-tech activity outside of their universities, below average wage levels, and high concentrations of employment in declining or stagnant industry sectors. Moreover, each region was facing a serious "brain-drain" problem, with large portions of the best and brightest graduates of the regions' universities leaving their respective regions because of a lack of job and career opportunities, particularly for those trained in science, engineering, and other technical fields. Business, political, and academic leaders in both regions agreed that intervention was necessary to reverse economic decline, diversify the economic bases, and bring high-paying jobs to their states.

In each case, an implicit theory of regional high-tech development was followed. In the Raleigh-Durham region, a blue-ribbon committee appointed by then Governor Luther Hodges was formed to consider what kind of strategy to pursue (Little, 1988). It was apparent to the committee that the region's greatest strength and principal comparative advantage, at least in the South, was the three research universities (the University of North Carolina at Chapel Hill, North Carolina State University, and Duke University). The combined research activity in the universities, the committee posited, was sufficiently large to attract a cluster of industrial research labs to the region to take advantage of proximity to sources of research products that had commercial application possibilities and a cadre of advanced graduate students for employment. The clustering of industrial research labs in the region, in turn, should induce the location of advanced manufacturing facilities to adjacent, but lower cost parts of the state. This theory was closely akin to growth pole, or growth center, doctrine; the "pole," however, was not a propulsive manufacturing industry but higher education.

The development practice was centered on the creation of a research park. The principal activity was recruitment, where the target was the R&D branch plants of major national corporations and federal government agencies. The technology areas targeted were those in which there was academic strength among the three research universities, namely, textile and organic chemistry, the environmental sciences, and, later, biotechnology and microelectronics. For the most part, the programs and incentives offered by state and local governments were nonpecuniary: the supply of infrastructure, favorable land use policies, specialized training programs through the new community college system, and access to the public universities. Later, several specialized state-sponsored research facilities—the Microelectronics Center of North Carolina and the Biotechnology Center—were developed to help promote industrial expansion in those technology areas within the state.

In the Salt Lake City region, development efforts were based on implicit theories of entrepreneurialism and regional comparative advantage. As was cited earlier, the region's brain-drain problem was well known. Local leaders assumed that, if job opportunities could be increased in science- and technology-based industries, then much of the talent leaving the region would stay because of the strong pull of the dominant Mormon culture for members of the Mormon faith. Moreover, the leaders felt that many of the highly educated people who had left the area for a richer set of career opportunities would readily return, if there were jobs commensurate with their skills.

Like the Raleigh-Durham region, the centerpiece of the strategy was the creation of a research park. But, unlike North Carolina, the policy approach was principally one of incubation of new technology-based businesses, in large part because entrepreneurialism was a strong element of Mormon culture. A wide variety of programs and investments were used, including land and buildings for incubator facilities, technical assistance,

technology transfer assistance, financing, and faculty incentives, by a combination of local and state governments. In addition, officials of the University
of Utah actively promoted "academic capitalism" among faculty and staff.

In both the Raleigh-Durham and the Salt Lake City cases, little high-
tech development occurred at first. In North Carolina, the trajectory
changed dramatically in 1965, about five years after the Research Triangle
Park was started, when R&D branch plants of IBM and the National
Institutes of Health were successfully recruited to the research park. Those
facilities served as "anchors" that helped attract other R&D branch plants,
mostly of multinational corporations.

There was also a lag between the creation of the University of Utah
Research Park and the changing fortunes of the Salt Lake City economy.
Two University of Utah faculty members who had started their own
businesses as outgrowths of their university-based research became highly
successful. One of the faculty members started spinning off new businesses from his original venture. Those became successful and in turn
parented a new generation of spin-offs. The other faculty member's company grew into one of the world's leaders in computer simulation and
imaging—Evans and Sutherland, Inc. These successes became entrepreneurial role models for others in the region to emulate.

Although quite different, the high-tech development strategies used in
the Research Triangle and Salt Lake City regions were each highly successful using any reasonable criterion of success. The key point for us is
that the strategies followed in North Carolina and Utah were consistent
with theories that were appropriate for those places and times.

The Need for a Contingent Theory of
Regional High-Tech Development

One way to confirm the appropriateness of underlying theory is to ask
whether the high-tech development policy approach used in Salt Lake City
would have been successful in North Carolina, and whether the approach
used in North Carolina would have worked in Utah (see Goldstein, 1991;
Luger and Goldstein, 1991). Our answer is no, for the simple reason that
starting regional conditions matter. In the case of Salt Lake City, a growth
center strategy based on recruiting R&D branch plants would probably not
have worked in the 1970s because it was widely perceived that non-
Mormon scientists, engineers, and other highly skilled workers would not
be willing to locate in a remote region whose culture and politics were
dominated by an unorthodox religious group. In the case of the Research

Triangle of North Carolina, there was little or no tradition of technology-based entrepreneurial activity, an underdeveloped business service infrastructure, and a lack of a skilled and well-educated resident labor force, all of which are needed to support an indigenous high-tech development strategy.

We cannot say whether the positive outcomes in Utah and North Carolina resulted from a conscious recognition by policymakers that initial conditions matter or whether happenstance played a role. The cases illustrate, nonetheless, the potential strength of a contingent approach, in which the local context, as well as theory, guide the selection of economic development programs.

Progress on the development of contingent theory(ies) of high-tech regional development rests on at least two fronts. The first lies in improved theory building. Existing theories need to be clarified and specified so that policies can be logically deduced from them. In other words, we should insist that the theories we publish in the scholarly literature be "empirically testable." The second thrust, an inductive approach, is systematic to undertake evaluations of specific regional high-tech policies and programs. When conducted across a broad range of regions and temporal periods, we can build up an empirical base over time for testing contingent theories of how regions develop technologically and what policies "work" under what conditions. Until then, we believe that theory and practice will continue to be separated by a wide gulf.

Notes

1. For definitions and discussions of technically sophisticated activities, process and product innovation, invention, and knowledge creation, see Stoneman (1983), Markusen et al. (1986), Luger et al. (1991), and Dosi and Orsenigo (1988).

2. We discuss another type of diffusion theory below: hierarchical diffusion, described in the economic geography and regional science, as opposed to the economics, literature.

3. *Preconditions* in this context refers to social and political factors, as well as economic ones, making it broader than its usage in neoclassical economics.

4. This material is drawn from Chapter 2 of Luger and Goldstein (1991).

5. See Holland (1976), Rees and Stafford (1985), and Gore (1984). Gore distinguishes Schumpeterian theories that "focus on the way in which entrepreneurial behavior and the act of innovation affect the dynamics of capitalist economies" (p. 86) from Perrouxian theories that stress "a range of mechanisms through which a propulsive unit could induce growth in many other parts of the economy" (p. 86).

6. See Krist (1985: 178-189). Allen and Hayward (1990) provide a more complete discussion of the dichotomy between regional economic (exogenous) and entrepreneurial (indigenous) approaches.

7. This taxonomy is similar to several others in the literature. See, for example, Luger (1987).

References

Aglietta, Michael (1976). *A Theory of Capitalist Regulation: The U.S. Experience.* London: NLB.

Allen, David N. and David J. Hayward (1990). "The Role of New Venture Formation/Entrepreneurship in Regional Economic Development." *Economic Development Quarterly* 4 (February): 55-63.

Andersson, Ake (1985). "Creativity and Regional Development." *Papers of the Regional Science Association* 56: 5-20.

Armington, Catherine, C. Harris, and M. Odle (1983). *Formation and Growth of High Technology Establishments: A Regional Assessment* (Grant No. ISI 82-12970). Final Report to the National Science Foundation, Washington, DC.

Arrow, Kenneth (1962). "Economic Welfare and the Allocation of Resources for Innovations" in R. Nelson (ed.), *The Rate and Direction of Innovative Activity* (pp. 609-625). Princeton, NJ: Princeton University Press.

Beckmann, M. and J. Thisse (1986). "The Location of Productive Activities" in P. Nijkamp (ed.), *Handbook of Regional and Urban Economics* (Vol. 1; pp. 21-95). Amsterdam: North-Holland.

Berry, Brian J. L. (1972). "Hierarchical Diffusion: The Basis of Development Filtering and Spread in a System of Growth Centers" in Niles Hansen (ed.), *Growth Centers in Regional Economic Development* (pp. 108-138). New York: Free Press.

Berry, Brian J. L. (1973). *Growth Centers in the American Urban System.* Cambridge, MA: Ballinger.

Bollinger, Lynn, Katherine Hope, and James Utterback (1983). "A Review of Literature and Hypotheses on New Technology-Based Firms." *Research Policy* 12: 1-14.

Boudeville, J. R. (1966). *Problems of Regional Economic Planning.* Edinburgh: Edinburgh University Press.

Britton, J. N. H. (1974). "Environmental Adaptation of Industrial Plants: Service Linkages, Locational Environment and Organization" in F. E. I. Hamilton (ed.), *Spatial Perspectives on Industrial Organization and Decision-Making* (pp. 363-390). New York: John Wiley.

Chow, G. C. (1967). "Technological Change and the Demand for Computers." *American Economic Review* 57: 1117-1130.

Clark, Gordon, Meric Gertler, and J. Whiteman (1986). *Regional Dynamics Studies in Adjustment Theory.* Winchester, MA: Allen and Unwin.

Cooper, A. and J. Komives (1972). *Entrepreneurship: A Symposium.* Milwaukee, WI: Center for Venture Management.

Cox, R. N. (1985). "Lessons from 30 Years of Science Parks in the U.S.A." in J. M. Gibb (ed.), *Science Parks and Innovation Centres: Their Economic and Social Impact* (pp. 17-25). Amsterdam: Elsevier Science.

Darwent, D. F. (1969). "Growth Poles and Growth Centers in Regional Planning: A Review." *Environment and Planning* 1 (1): 5-31.

Dasgupta, P. and J. Stiglitz (1980). "Industrial Structure and the Nature of Innovative Activity." *Economic Journal* 90: 226-293.

Dixon, R. (1980). "Hybrid Corn Revisited." *Econometrica* 48: 1451-1462.

Dosi, G. and L. Orsenigo (1988). "Coordination and Transformation: An Overview of Structures, Behaviors and Change in Evolutionary Environments" in Giovanni Dosi et al. (eds.), *Technical Change and Economic Theory* (pp. 13-37). London: Frances Pinter.

Erickson, Rodney A. (1975). "The Spatial Pattern of Income Generation in Lead Firm, Growth Area Linkage Systems." *Economic Geography* 51: 17-26.

Freeman, C., J. Clark, and L. Soete (1982). *Unemployment and Technical Innovation: A Study of Long Waves and Economic Development*. London: Frances Pinter.

Gilmour, J. M. (1974). "External Economies of Scale, Interindustry Linkages and Decision-Making in Manufacturing" in F. E. I. Hamilton (ed.), *Spatial Perspectives on Industrial Organization and Decision-Making* (pp. 335-362). New York: John Wiley.

Glasmeier, Amy (1990). *The Making of High Tech Regions*. Princeton, NJ: Princeton University Press.

Goldstein, Harvey (1991). "Growth Centers vs. Endogenous Development Strategies: The Case of Research Parks" in E. Bergman, G. Maier, and F. Tödtling (eds.), *Regions Reconsidered*. London: Manasell.

Gore, Charles (1984). *Regions in Question: Space, Development Theory and Development Policy*. London: Methuen.

Griliches, Z. (1957). "Hybrid Corn: An Exploration in the Economics of Technical Change." *Econometrica* 25: 501-522.

Hansen, Niles (1967). "Development of Pole Theory in a Regional Context." *Kyklos* 20: 709-725.

Harris, Candee S. (1986). "Establishing High-Technology Enterprises in Metropolitan Areas" in Edward M. Bergman (ed.), *Local Economies in Transition* (pp. 165-184). Durham, NC: Duke University Press.

Harvey, David (1982). *The Limits to Capital*. Chicago: University of Chicago Press.

Higgins, Benjamin (1983). "From Growth Poles to Systems of Interaction in Space." *Growth and Change* (October): 7.

Hirschman, Albert O. (1958). *The Strategy of Economic Development*. New Haven, CT: Yale University Press.

Holland, Stuart (1976). *Capital Versus the Regions*. New York: St. Martin's.

Isard, W. (1956). *Location and Space Economy*. Cambridge: MIT Press.

Krenz, Charles (1987). "Silicon Valley Spin-Offs." Unpublished master's thesis, Stanford University Business School.

Krist, H. (1985). "Innovation Centres as an Element of Strategies for Endogenous Regional Development" in J. M. Gibb (ed.), *Science Parks and Innovation Centres: Their Economic and Social Impact* (pp. 178-189). Amsterdam: Elsevier Science.

Lasuén, J. R. (1971). "Multi-Regional Economic Development: The Open System Approach" in T. Hagerstrand and A. Kuklinski (eds.), *Information Systems for Regional Development: A Seminar* (Lund Studies in Geography, series B, Human Geography 37, pp. 169-212). Lund: Gleerup.

Lasuén, J. R. (1973). "Urbanization and Development: The Temporal Interaction Between Geographical and Sectoral Clusters." *Urban Studies* 10: 163-188.

Little, William F. (1988). "Research Triangle Park." *The World and I* (November): 178-185.

Luger, Michael I. (1987). "State Subsidies for Industrial Development: Program Mix and Policy Effectiveness" in John M. Quigley (ed.), *Perspectives on Local Public Finance and Public Policy* (pp. 29-61). Greenwich, CT: JAI.

Luger, Michael I. and Harvey A. Goldstein (1991). *Technology in the Garden, Research Parks and Regional Economic Development*. Chapel Hill: University of North Carolina Press.

Luger, Michael I., Uwe Schubert, and Franz Tödtling (1991). *External and Internal Factors Affecting the Productivity of European Research Institutes*. Vienna, Austria: Wirtschaftsuniversitat Wien, IIR (October).

Malecki, Edward and Pravin Varaiya (1986). "Innovation and Change in Regional Structure" in Peter Nijkamp (ed.), *Handbook of Regional and Urban Economics, Vol. 1: Regional Economics* (pp. 629-645). Amsterdam: North-Holland.

Mansfield, E. (1968). *Industrial Research and Technological Innovation*. New York: Norton.

Markusen, A. (1987). *Profit Cycles, Oligopoly, and Regional Development*. Cambridge: MIT Press.

Massey, Doreen (1984). *Spatial Divisions of Labour: Social Structures and the Geography of Production*. London: Macmillan.

Mensch, G. (1979). *Stalemate in Technology: Innovations Overcome the Depression*. New York: Ballinger.

Moore, Mark (1980). "Social Science and Policy Analysis, Some Fundamental Differences." Cambridge, MA: John F. Kennedy School of Government.

Moseley, Malcolm J. (1975). *Growth Centers in Spatial Planning*. Oxford: Pergamon.

Moulaert, Frank and Eric Swyngedouw (1991). "Regional Development and the Geography of Flexible Production Systems" in Ulrich Hilpert (ed.), *Regional Innovation and Decentralization: High Tech Industry and Government Policy* (pp. 239-267). London: Routledge.

Myrdal, Gunnar (1957). *Economic Theory in Underdeveloped Regions*. London: Duckworth.

Nabseth, L. and G. F. Ray (1974). *The Diffusion of New Industrial Processes: An International Study*. Cambridge: Cambridge University Press.

Ohlin, B. (1933). *Interregional and International Trade*. Cambridge, MA: Harvard University Press.

Perroux, Francois (1950). "Economic Space, Theory, and Applications." *Quarterly Journal of Economics* 64: 89-104.

Pred, Allen (1976). "The Interurban Transmission of Growth in Advanced Economies: Empirical Findings Versus Regional Planning Assumptions." *Regional Studies* 10: 151-171.

Rees, John, R. Briggs, and D. Hicks (1985). "New Technology in the United States' Machinery Industry" in A. T. Thwaites and R. P. Oakley (eds.), *The Regional Economic Impact of Technological Change* (pp. 164-194). London: Frances Pinter.

Rees, John and Howard Stafford (1985). "A Review of Regional Growth and Industrial Location Theory: Towards Understanding the Development of High-Technology Complexes in the United States." Paper prepared for the Office of Technology Assessment, U.S. Congress.

Rostow, Walt W. (1991). *The Stages of Economic Growth: A Non-Communist Manifesto* (3rd ed.). Cambridge: Cambridge University Press.

Sabel, Charles (1982). "The End of Fordism?" Chapter 5 in *Work and Politics*. Cambridge: Cambridge University Press.

Salter, W. E. G. (1966). *Productivity and Technical Change* (2nd ed.). Cambridge: Cambridge University Press.

Scherer, F. M. (1965). "Firm Size, Market Structure, Opportunity and the Output of Patented Inventions." *American Economic Review* 55: 1103-1113.

Schumpeter, J. A. (1939). *Business Cycles: A Theoretical, Historical, and Statistical Analysis of the Capitalist Process*. London: McGraw-Hill.

Scott, A. and M. Storper (1987). "High Technology Industry and Regional Development: A Theoretical Critique and Reconstruction." *International Social Science Journal* 39 (12): 215-232.

Shapero, Albert (1981). "Entrepreneurship: Key to Self-Renewing Economies." *Commentary* 5 (April): 19-23.

Stoneman, P. (1981). "Intra Firm Diffusion, 'Baysian' Learning and Profitability." *Economic Journal* 91: 375-388.

Stoneman, Paul (1983). *The Economic Analysis of Technological Change*. Oxford: Oxford University Press.

Tayler, C. and Z. A. Silberston (1973). *The Economic Impact of the Patent System*. Oxford: Oxford University Press.

Teece, D. J. (1977). "Technological Transfer of Multinational Firms: The Resource Cost of Transferring Technological Know How." *Economic Journal* 87: 242-261.

Thomas, M. D. (1985). "Regional Economic Development and the Role of Innovation and Technical Change" in A. T. Thwaites and R. P. Oakley (eds.), *The Regional Economic Impact of Technological Change* (pp. 13-39). London: Francis Pinter.

Thwaites, A. T., A. Edwards, and D. Gibbs (1982). *Interregional Diffusion of Product Innovations in Great Britain* (Final Report to the Department of Industry and the EEC). Newcastle: Centre for Urban and Regional Development, Newcastle University.

Thwaites, A. T., R. P. Oakey, and P. A. Nash (1982). "Technological Change and Regional Economic Development: Some Evidence on Regional Variation in Product and Process Innovation." *Environment and Planning A* 14: 1073-1086.

U.S. Congress, Office of Technology Assessment (1985). *Technology, Innovation, and Regional Economic Development, Background Paper 2: Encouraging High-Technology Development.* Washington, DC: Office of Technology Assessment, U.S. Congress (February).

Vernon, R. (1966). "International Investment and International Trade in the Product Cycle." *Quarterly Journal of Economics* 80: 190-207.

Vietorisz, Thomas and Bennett Harrison (1973). "Labor Market Segmentation: Positive Feedback and Divergent Development." *American Economic Review* 63: 366-376.

Weber, Alfred (1929). *Theory of the Location of Industry* (translated by C. Friedrich). Chicago: University of Chicago Press.

Suggested Readings

Clark, John, Christopher Freeman, and Luc Soete (1983). "Long Waves, Inventions, and Innovations" in Christopher Freeman (ed.), *Long Waves in the World Economy* (pp. 63-77). London: Butterworths.

Freeman, Christopher (1974). *The Economics of Industrial Innovation.* Harmondsworth: Penguin.

Glasmeier, Amy (1990). *The Making of High Tech Regions.* Princeton, NJ: Princeton University Press.

Goldstein, Harvey (1991). "Growth Centers vs. Endogenous Development Strategies: The Case of Research Parks" in E. Bergman, G. Maier, and F. Tödtling (eds.), *Regions Reconsidered.* London: Manasell.

Gore, Charles (1984). *Regions in Question: Space, Development Theory and Development Policy.* London: Methuen.

Kondratiev, N. D. (1935). "The Long Waves in Economic Life." *Review of Economic Statistics* 17: 149-158.

Luger, Michael I. (1987). "State Subsidies for Industrial Development: Program Mix and Policy Effectiveness" in John M. Quigley (ed.), *Perspectives on Local Public Finance and Public Policy* (pp. 29-61). Greenwich, CT: JAI.

Malecki, Edward J. (1991). *Technology and Economic Development: The Dynamics of Local, Regional, and National Change.* New York: John Wiley.

Markusen, Ann, Peter Hall, and Amy Glasmeier (1986). *High Tech America: The What, How, Where and Why of the Sunrise Industries.* Boston: Allen and Unwin.

PART IV

Political and Social Theories

8

Political Economy and
Urban Development

C. SCOTT HOLUPKA
ANNE B. SHLAY

The 1992 Los Angeles riots have once again brought issues of urban inequalities to national attention. The policy debate following the riots, however, illustrates how different premises about the causes of the unrest lead to very different policy considerations. Conservatives, seeing the unrest as resulting from a lack of personal pride and inadequate investment opportunities, prescribed home ownership and free enterprise zones. Alternatively, more liberal policymakers emphasized the decline in government spending on social programs, education, and job development and consequently suggested greater spending and more direct aid to cities.

These different policy suggestions underscore an insight familiar to anyone who has thought about or been involved in public policy decision making: Deciding whether, when, and how to intervene in a process is predicated upon how the process is understood to operate. This insight, in turn, highlights the importance of theory to public policymaking, because the question of how a social process works is always, at heart, a theoretical matter. Different policy suggestions often reflect different theoretical views about a process. Likewise, different theoretical perspectives can yield radically different policy options and ideas.

The theoretical perspective upon which U.S. urban development policies have traditionally been based maintains that the structure of the urban landscape is a result of a free market where firms and individuals compete for space. Urban development trends are the product of macro forces emerging from the micro decisions of people, not conscious design or plans. Following this perspective, land use planning, such as zoning, is not a tool to alter development trends but to prevent certain negative externalities, to maximize land values, and to promote development at its "highest and best use." The implications of this outlook are important. If urban

175

development patterns emerge from the decisions of the many, they can neither be controlled nor altered. Moreover, a free market perspective implies that intervention in the urban development process is not desirable because it is assumed that free markets operate optimally.

Other theoretical perspectives of urban development are being developed, however, that take a very different view of urban development. One of the major, evolving new perspectives of urban development, variously described as *political economy theory,* the *institutional approach,* or the *new urban theory,* emphasizes how the actions and behaviors of key institutions and actors are responsible for determining how urban structures develop. Urban development outcomes viewed as "inevitable" under the traditional perspective are seen as being contingent upon specific actions taken by specific actors and institutions in this new theoretical framework. By arguing that the existing urban structure, with all its inequities, is not inevitable, this new theoretical perspective suggests new, now-possible public policy initiatives and opportunities.

This chapter examines how a political economic perspective on urban development can be applied to help develop more effective as well as progressive public policy interventions that can alter development trends and redress some urban inequities.

Yet, taking on a political economic perspective has implications that are broader than pointing toward selected targets for intervention (e.g., the downtown big project) or directions for public policy (e.g., equity planning). It also has implications for the political strategies and tactics used to alter the course of development. "Applying" a political economic perspective is not a linear process but requires a clear recognition of the political dynamics and structures of power that are central to the way development occurs. Intervening in metropolitan spatial dynamics using a political economic perspective requires a willingness to engage in political struggle with powerful people, institutions, and organizations. Altering the course of metropolitan development means interfering with the distribution of benefits associated with established and taken-for-granted patterns. Therefore intervening in these patterns will not go unnoticed because, for the public to win, someone has to lose.

The chapter introduces this new perspective on urban development as a tool for economic development activities. It begins by outlining its key points, focusing particularly on the elements and workings of a "growth machine." The second part describes some empirical illustrations of cases where applications of this perspective appear to be working to begin to alter destructive paths of development. These cases include an examination of efforts to halt, redirect, or extract larger benefits from downtown projects and recent political activities around the investment activities of financial institutions.

Political Economic Theory:
An Overview

Historically, the dominant models of urban development have assumed that the "invisible hand" of the marketplace guides the placement of industrial, commercial, and residential uses. Early examinations of urban development in sociology and economics viewed the impersonal market competition for sites, as expressed through land values, rents, and the like, as analogous to the biotic competition between various species of plants and animals for a niche in a particular environment. Just as competition in nature results in the best suited plants and animals surviving, it was assumed that a given pattern of urban development represents the best use of the urban environment.[1]

It has been suggested that the free market reasoning that forms the basis of this ecological perspective has been so deeply ingrained in many scholars that they "seem not to have been aware that there was even an alternative approach" (Logan and Molotch, 1987: 4). Others have suggested that important insights from the ecological school, such as the invasion-succession model, have become so taken for granted that they have been reified into social laws.

For example, Bradford has suggested that the widespread adherence in the real estate industry to these early models of neighborhood change creates a self-fulfilling prophecy where the assumption of neighborhood decline led real estate institutions to make decisions that, in fact, foster neighborhood decline (Bradford, 1976). In this case, theory in the form of urban ecology actually helps shape urban reality.

In response to the deleterious consequences and political immobilization stemming from urban ecology as a reigning paradigm, a set of recent approaches have been brought together to inform a new, developing perspective on urban development. This new perspective began as a critique of urban ecology but is now starting to make contributions that define it as a clear alternative to ecological modes of thought. It has been variously described as the *new urban sociology* (Zukin, 1980; Gottdiener and Feagin, 1988), the *new urban politics* (Stone, 1991), the *institutional approach* (Shlay, forthcoming), or *political economic theory* (Molotch, 1976).

This emerging framework for understanding urban development has several premises. The first is the recognition that a constellation of specific local actors and institutions play key roles in determining how local development occurs; metropolitan patterns are not emergent phenomena from the micro decisions of the many but are directed by decisions internal to local organizations. The second is that the distribution of benefits from development are highly skewed; those at the top reap the benefits from development while the public largely pays the costs of development. The third premise is that metropolitan development decision-making processes tend to be undemocratic; critical decisions are typically hidden from the public view and the public serves to legitimize decisions rather than to play a material role in making them.

A political economic perspective is rooted in Marxist theory (Harvey, 1973). Although Marx never focused on urban development as a central part of his theory, later scholars explicitly worked to develop a Marxist theory of cities and metropolitan development (Harvey, 1973; Castells, 1976; Smith, 1986; Walker, 1981). Framed in this way, metropolitan patterns represented the physical expression of a mode of production—capitalism, where urban form reflects system requisites for capitalist accumulation. Central urban concepts such as location, rent, central place, and scarcity are explained in class terms—as the outcome of struggle and negotiations between classes, not the equilibrium point between supply and demand functions. With capitalism as the explanatory underpinnings for urban form, neoclassical economic theory and urban ecology are not shown to be lawlike but to be historically specific. As noted in the early work of David Harvey:

> All land use theories must be regarded as contingent. They are only specific theories which can play specific roles in helping to elucidate a particular set of assumptions about the dominant mode of production, about the nature of social relationships, and under the prevalent institutions of society. (1973: 192)

Indeed, treating urban ecology as lawlike makes it part of a superstructure that helps to preserve the capitalist mode of production.

A Marxist theory of the city does not describe an open, fluid market but begins to show how various forces and class relationships in the political and economic spheres constrain locational decisions. Moreover, Marxism provides a political context for thinking about who wins and loses in struggles on the urban scene and why. Power and politics are an explicit part of a Marxist theory of the city.

Yet a major objection to Marxist analyses is that, like ecological perspectives, the emphasis on structural forces tends to neglect or ignore the importance of individual action (Logan and Molotch, 1987; Feagin, 1988; Gottdiener and Feagin, 1988). The determinism embodied in explanations like "requirements of capitalist accumulation" and other economic forces emanating from the struggle between capital and labor renders intervention virtually impossible and "avoids working through how human activities give social structures their reality" (Logan and Molotch, 1987: 10).

The importance of understanding how specific actors and activities influence urban development was first proposed by William Form (1954), who identified four sets of associations and interests—real estate and building businesses, large businesses and utilities, home owners and small land owners, and local governmental agencies—as being important for determining land use decisions and maintaining the appearance of an ecological order. But the role of politics and agency in shaping place was given its clearest contemporary expression with the recognition that the city was a "growth machine" (Molotch, 1976).

A growth machine perspective argues that city development represents the collective and concerted activities of growth coalitions who deliberately work to develop and change the urban landscape. With the goal of intensifying land use, increasing the local population base, and, ultimately, enlarging the demand for local goods and services, urban growth coalitions are the agents behind development on the urban scene. Either visibly leading the way or, more typically, out of sight, city growth coalitions are viewed as the source of energy and direction for the development schemes that count (Molotch, 1976, 1979; Molotch and Logan, 1984, 1985; Logan and Molotch, 1987).

The machinelike quality of the workings of growth coalitions is due to their ability to legitimize their authority as "civic" leaders as well as the automatic acceptance of the goodness of growth by the public. Articulating that growth is universally beneficial and, indeed, that its absence would be costly for all, growth coalitions are so much a part of a metropolitan development praxis that a progrowth agenda is accepted as common sense. Hence the city becomes a growth machine by virtue of a local growth coalition's ability to conduct business in such a way that alternative courses are inconceivable.

Growth coalitions represent a constellation of interests most concerned with land development and metropolitan growth. Key players out front are politicians, the local media, developers, financial institutions, and utility companies. Providing much of the auxiliary support for the growth machine are universities, cultural institutions (e.g., museums), professional sports, and unions (Logan and Molotch, 1987).[2] Shared networks (e.g., clubs, corporate boards, committees, and civic organizations) and shared interests provide the opportunity and motivation for collective action.

The success of local growth machines depends on their ability to remain unaccountable while maintaining an appearance of openness in the politics that guides development. With the prerogatives of growth dutifully promulgated by the media, particularly by newspapers, growth coalitions are able to orchestrate the course of urban development.

The facade of democracy is necessary because the true benefits and costs are hidden from the public domain. Development projects are "sold" to communities and often provided with large public subsidies with the rationale that development will benefit the community—the promised economic spillovers of the big project. But the benefits of growth are almost always skewed in favor of landowning elites. Moreover, growth appears to be accompanied by many negatives—higher housing costs and problems of housing affordability, increased congestion and pollution, greater demand for public services, and higher taxes. Indeed, it is not certain that growth per se results in increased employment although it is often promoted for that reason. Therefore authentic democracy and a true accounting of the costs and benefits associated with development in informing decisions over the direction of urban development surely might undermine growth machine endeavors.

Yet it is also important to recognize that growth coalitions are not free agents in setting development agendas. The success of a growth coalition depends on its ability to attract investment from national and international markets. Because growth machines are so dependent on outside capital, they must be attentive to the needs of capital. Therefore growth machine actors are essentially intermediary actors in a much larger capitalist system (Molotch, 1979).

The city as a growth machine thesis pulls together several important strands of urban research. It incorporates notions of structural constraint on local action, particularly constraints imposed by an international, capitalist economy, while still allowing for local variation and individual action. This new perspective is thus fusing structuralist arguments with a more action-oriented view of urban development (Gottdiener and Feagin, 1988).[3] A major task for urban research is identifying which institutions and actors are particularly powerful in affecting urban development.

In developing a better understanding of which actors and institutions are important for determining how the urban landscape changes, this new perspective also provides a new model for directing urban policy and advocacy. By arguing that a particular change in the urban landscape is neither inevitable nor natural, but instead is a result of specific decisions made by specific actors and institutions, this new urban theory both encourages and guides attempts to intervene in the development process.

Fighting Against Inequities in
Urban Development: Case Studies

Despite its recent development, the emerging political economy or institutional perspective has been used by some individuals and organizations involved in altering the consequences of current urban development practices. A review of some of these efforts shows how this new urban theory has helped people decide where and how to intervene in the development process to create a more equitable urban structure.

Attacking the Big Project

A hallmark of urban development since the 1950s has been the large renewal project. Involving large tracts of land in prime city locations, together with large sums of money to finance them, major redevelopment projects have transformed the downtowns of most of the nation's cities.

Support for these major development efforts has usually been provided by a consistent coalition of actors and institutions, the ones referred to as members of the growth machine. These include large local financial institutions, developers and real estate officials, utility company representatives, and the local media. In many cities, the activities of these various interests have been coordinated through a committee established precisely to direct growth. Well-known examples include Pittsburgh's Allegheny Conference on Community Development, the "Vault" in Boston, and the Greater Baltimore Committee. These groups are often credited with guiding their respective cities from a manufacturing-based economy to service economy cities.

Another key player in these development efforts has been the government. Local government, in particular, has usually been called upon to help assemble required parcels of land, to provide major infrastructure improvements such as new roads and sewers, and often to subsidize the development either directly with loans or loan guarantees or indirectly with tax abatements and other write-offs.

Yet in recent years some efforts by local land-based elites to initiate massive renewal projects have been met with resistance. A few projects have been completely stopped, while others have been modified or additional concessions have been exacted before the projects have been allowed to proceed. Those who have attacked these big projects have been aided by the development of the new urban theory.

One of the best examples of fighting the big project by taking on the growth machine is the successful effort to halt a world's fair in Chicago for 1992 (Shlay and Giloth, 1987; McClory, 1986). Since the 1960s, several U.S. cities have used world's fairs as tools for downtown expansion and urban redevelopment. As a vehicle for urban renewal, fair preparations have enabled downtown land acquisition and clearance; have built commercial structures, convention centers, opera houses, and hotels; and have created parks, landscaping, and improved downtown highway transportation. These represent the "residuals" from hosting a world's fair.

Like other big projects, world's fair boosters promoted these enormous six-month events as economic development projects, enlisting the support of the government, the public, and, importantly, public coffers. Public dollars represented the bulk of the financing for the modern world's fair. Yet, out of the six fairs held since 1960, only one showed a profit while another broke even. Losses were passed on to stockholders, consumers, and taxpayers.

Chicago's world's fair was to be located at the southern edge of its downtown, the Loop. The selected site required massive landfill as well as major transportation modifications. Estimated costs for the fair peaked at $1 billion. Promised spillovers from the fair included new jobs, increased revenues, and the renewal and redevelopment of Chicago's south side. Fair benefits, it was assumed, would far outweigh its costs and Chicago would then join the ranks of other world class cities.

Leadership for support for the fair came from the chief executive officer of Chicago's largest utility company, Commonwealth Edison. With other fair backers, he helped to form a coalition that became officially organized as a nonprofit organization—the 1992 Fair Corporation. The task of the 1992 Fair Corporation was to make the 1992 fair happen in Chicago. Initially so successful in promoting this effort, the fair was incorporated as part of Chicago's 1982 master plan for guiding downtown development. Supported by the mayoral administration and with the requisite national and international designations, a world's fair seemed inevitable for Chicago.

The 1992 Fair Corporation was the classic manifestation of an urban growth machine. Board membership was dominated by representatives of manufacturing, publishing, communication, advertising, energy, banking and insurance, and development. Research revealed a web of relationships among board members through corporate interlocks (Shlay and Giloth, 1985a, 1985b). These people were also shown to be already working together in business and planning organizations that support corporate interests. Moreover, this group was shown to be a social community as exhibited by the network of corporate memberships in elite clubs (Shlay and Giloth, 1987). A highly organized and cohesive group of powerful people with huge resources were able to set Chicago's development agenda.

Yet by 1985 fair plans had been put to rest and the city abandoned this project. Launched with civic unanimity, the 1992 fair was toppled through sophisticated organizing that incorporated a political economic perspective and the new urban theory.

In 1981 a group began meeting over concerns about how a world's fair would affect Chicago, particularly its working-class neighborhoods and communities of color (McClory, 1986). Soon to be called the Chicago 1992 Committee, this group initiated an organizing effort representing 48 separate community organizations that would culminate in the cessation of fair planning.

Two key elements of the new urban theory were important in organizing against the fair. First, the Chicago 1992 Committee recognized that planning a world's fair for Chicago was not democratic and questioned the legitimacy of the 1992 Fair Corporation to lead this process. It began establishing a discourse where 1992 Fair Corporation members were not articulated as civic leaders but as business people with interests.

Second, it raised questions about the prerogatives of growth and questioned accepted conventions about the costs and benefits from development. Following the perspective that fair planners had economic interests in the fair, organizers began to investigate the costs and benefits associated with the fair. Commissioning studies and conducting their own research, they began to hammer away at the assumption that a fair would help Chicago, particularly its most disadvantaged citizens (Weiss and Metzger, 1989). To the contrary, they began to uncover a pattern of sloppy planning, poor research, and general incompetence by people at the heart of the fair planning process (Shlay and Giloth, 1987; McClory, 1986).

But democracy itself was decisive in toppling the fair, a change in Chicago politics that permitted organizations like the Chicago 1992 Committee to ask questions, acquire resources, and mobilize attention on the activities of Chicago's growth coalition (Shlay and Giloth, 1987). This democratic process was initiated under the Harold Washington administration, which, unlike the previous regime, was not a fair booster. This set the political context for raising the fundamental issues about costs and benefits that ultimately led to the fair's demise.

Benefiting from the kinds of analyses that the new urban theory emphasizes, the community opposition was able to show that potential benefits of the fair were neither as large nor as evenly distributed as its proponents had said, while the costs, both in tax-supported subsidies and in neighborhood displacement, would be quite high (Weiss and Metzger, 1989). As the public risks and private rewards became more widely known, public support for the fair disappeared.

Completely stopping redevelopment efforts has not usually been possible, or even necessarily desired, elsewhere in the country. But in a number of cases, community groups and local politicians have been able to extract concessions or trade-offs from developers in exchange for allowing development efforts. In Boston, for instance, developers have been required to pay special assessments beyond the normal permit and development fees in order to operate in the city. These special assessments have been used to fund construction and rehabilitation of low- and moderate-income housing elsewhere in the city (Dreier, 1989). Boston's Flynn administration has also pushed for inclusionary zoning practices, requiring developers to set aside some housing units for lower income residents.

Local efforts restricting growth have also been successful in varying degrees in a number of California cities. San Francisco has instituted linked development projects similar to Boston's. Santa Monica, with over a decade of progressive politicians in office, has enacted rent control ordinances and produced development agreements requiring developers to include art and social service fees, affordable housing, day-care facilities, and public space in any large project. And Santa Barbara residents successfully stopped offshore drilling proposals and have enacted strong environmental policies limiting growth in the area (Molotch, 1990; Logan and Molotch, 1987).

These successful efforts in stopping development projects, or reducing the inequities normally produced by them, show that urban development processes can be altered. With effort and persistence, development practices can be made to serve broader public interests. The developing political economic framework has helped in these efforts by (a) identifying the key actors who need to be influenced, (b) documenting the private benefits and public expenses often associated with these endeavors, and (c) raising awareness that no specific project or policy is either inevitable or natural.

Rethinking Home Financing

One of the best examples of how changes in our understanding of urban development can encourage new ways of intervening and modifying the development process has been in the field of mortgage lending and home financing. While some of the research on home financing can be traced to theoretical developments of the late 1960s and early 1970s—notably the works of political economic writers such as Harvey (1973, 1977, 1985; Harvey and Chatterjee, 1974)—most lending research can be traced to the efforts of community organizers and activists. Disheartened by deteriorating housing and communities, and outraged at the block-busting practices of real estate entrepreneurs, some community activists in the late 1960s began to argue that these conditions were related to mortgage lending practices. Challenging the traditional explanation of neighborhood decay as an inevitable and natural occurrence, community groups and activists were among the first to attempt to collect pertinent mortgage lending data.

Despite difficulties in obtaining information, a stable pattern of lending practices was found in city after city. Large disparities were found in the volume of lending made in suburbs compared with older, central-city neighborhoods. Studies also found substantially less lending in minority or racially changing neighborhoods.

To the growing body of empirical work demonstrating differences in lending, urban scholars began to add a theoretical framework. Among the most important theoretical models developed was Bradford and Rubinowitz's "Urban-Suburban Investment-Disinvestment Process" (Bradford and Rubinowitz, 1975). Bradford and Rubinowitz argued that urban decay was not the result of "natural" economic processes removed from the control of public policy, as traditional ecological and economic theory suggested. Instead, echoing Form's (1954) remarks, they said that urban decay and suburban growth were "the result of identifiable private and public investment decisions, made by identifiable public actors and members of the real estate investment and development industry" (Bradford and Rubinowitz, 1975: 424). Suburban investment and development, and urban disinvestment, they argued, were a single, interrelated process mediated by investors. The lending and investment decisions made by such key actors as real estate developers, banks, and savings and loans created a self-fulfilling prophecy of urban decay and suburban growth.

Bradford and Rubinowitz's "Investment-Disinvestment" model encouraged still further examinations of the activities of lending institutions. Reacting to this research, as well as to the political efforts of community groups, several states in the mid-1970s passed legislation requiring financial institutions to make public their lending records. It was felt that disclosure of such data would make it easier for communities to monitor the actions of individual lending institutions, thus making it more difficult for such institutions to engage in discriminatory and other forms of unfair lending practices.

The federal government was also urged to pass lending disclosure legislation, because most financial institutions were federally insured and regulated. The federal government response was the Home Mortgage Disclosure Act (HMDA) in 1975. HMDA requires all federally insured lenders with net assets of $10 million or more to report the number and amount of home mortgage and home improvement loans by census tract within their market areas. HMDA provides community groups, and urban scholars, with more readily available information on mortgage lending activity.

In 1977 Congress enacted the Community Reinvestment Act (CRA). This act states that lenders have an affirmative obligation to serve the credit needs of all communities within their local area, including minority and low- or moderate-income areas. Community groups have been able to use the CRA to gain concessions from individual lenders, using HMDA research to support claims of discriminatory lending practices.

In the 1980s computerized HMDA reports made lending information more readily available to scholars, starting a new round of lending research. Using the computerized data, and more sophisticated statistical modeling procedures, lending patterns in a number of cities across the nation have been analyzed (Goldstein, 1986; Shlay, 1988, 1987, 1986). Studies have consistently shown that nonmarket forces, particularly neighborhood racial composition, play a significant role in mortgage lending patterns. These studies have further fueled community reinvestment activity around the country. In city after city, community groups have obtained agreements from major lenders to directly increase investments in low-income and minority neighborhoods and to work jointly with other lenders, local government, and community-based organizations to develop loan pools and lending consortia that can make loans available to previously neglected communities.

Community reinvestment organizing, and research on home financing, have thus had a synergistic relationship (Shlay, 1989). Community activity has encouraged, sometimes even sponsored, research on home financing. The research, in turn, has encouraged and helped focus many community reinvestment efforts. More recent urban theories, in particular, have shown how lending decisions affect neighborhood development. While traditional explanations of urban change emphasized the "unseen," and arguably uncontrollable, hand of an open market, theories of urban change like Bradford and Rubinowitz's "Investment-Disinvestment" model provide a concrete explanation for neighborhood decay and deterioration. By focusing on the activities of banks and savings and loans, community groups are attempting to alter the conditions that produced earlier periods of decay and decline.[4]

Conclusion

Applying a political economy perspective to redressing inequities in urban development means taking account of the decisions and activities of urban institutions that govern spatial patterns. In particular, it requires making visible the workings of metropolitan growth coalitions that are the engineers of urban development. A first step toward intervening in the taken-for-granted development processes is determining the institutions, actors, and decisions that shape place.

Yet, as the empirical cases in this chapter suggest, actual intervention in development activities requires more than simply describing the actors that are behind seemingly uncontrollable urban forces. As the battle over Chicago's would-be world's fair shows, applying a political economy perspective means opening up decision-making processes and seeking a greater popular voice in development decisions that matter. For putting a political economy perspective to work on the urban scene is ultimately political because it is geared toward democratizing development processes. It is directed at creating an informed public who have available a complete account of the costs and benefits associated with development.

Yet, as shown by community struggles over lending practices and community reinvestment, using a political economy perspective is about making institutions accountable as well as law abiding. Much of the recent conflict over lender investment practices has been over whether lenders will be permitted to continue to disinvest from communities, discriminate against minority communities, and generally avoid obeying laws governing equal opportunity rights, community reinvestment obligations, and the like. In this sense, a political economy perspective leans toward efforts that promote fairness and justice.

New theoretical developments in the urban field indicate that features of the urban landscape that were previously thought to be "natural" or "inevitable" are now seen as being contingent upon specific actions made by specific actors and institutions. As a consequence, those with the courage to try to change the urban structure and to address current inequities are given hope that such efforts can be successful and provided with some guidance as to where efforts should be focused.

Further research is needed to better understand which actors and institutions are important in influencing urban development, how these actions and behaviors are constrained by broader social forces, and where other organizations, institutions, and individuals might best intervene to change some of the most damaging consequences of the current development process. To the extent that this research agenda is successful, practitioners will gain a better understanding of how more equitable urban development can be achieved.

Notes

1. See R. D. McKenzie's "The Ecological Approach to the Study of the Human Community" (1967) for a seminal discussion of this viewpoint.

2. Similarly, Kevin Cox has noted the importance of "locally dependent" firms in driving local economic development and politics (Cox, 1983; Cox and Mair, 1988). Although Cox's definition of "locally dependent" firms does not entirely overlap with Molotch's "growth coalition," both emphasize businesses whose profitability is linked to investment and growth within a specific geographic area.

3. In a similar vein, Stone's "regime theory" discusses how government plays a mediating role in linking market forces to policy outcomes (Stone, 1987, 1991). Like the growth machine model, regime theory recognizes that the capitalist market imposes constraints upon government officials but also maintains that there is still a great deal of variability in the policies governments pursue.

4. One particularly unusual case was where a working-class community in Philadelphia successfully battled in court a large lender seeking to close a local branch in the neighborhood.

With the argument that the branch closing would have severe consequences for the neighborhood, they were able to persuade the court to grant an injunction against the branch closing for a specific amount of time.

References

Bradford, C. (1976). "Real Estate Appraisal and Racial Discrimination: A Conspiracy of Beliefs." Statement prepared for a meeting of the Office of Fair Housing and Opportunity, U.S. Department of Housing and Urban Development (July 16).

Bradford, C. and L. Rubinowitz (1975). "The Urban-Suburban Investment Process: Consequences of Older Neighborhoods." *The Annals* 422: 77-86.

Castells, M. (1976). "Is There an Urban Sociology?" in C. G. Pickvance (ed.), *Urban Sociology* (pp. 33-59). London: Tavistock.

Cox, K. (1983). "Local Interests and Urban Political Processes in Market Societies" in R. Lake (ed.), *Readings in Urban Analysis* (pp. 103-117). New Brunswick, NJ: Rutgers University Press.

Cox, K. and A. Mair (1988). "Locality and Community in Politics of Local Economic Development." *Annals of the Association of American Geographers* 78: 307-325.

Dreier, P. (1989). "Economic Growth and Economic Justice in Boston: Populist Housing and Jobs Policies" in G. D. Squires (ed.), *Unequal Partnerships: Political Economy of Urban Redevelopment in Postwar America* (pp. 35-38). New Brunswick, NJ: Rutgers University Press.

Feagin, J. R. (1988). *Free Enterprise City: Housing in Political and Economic Perspective.* New Brunswick, NJ: Rutgers University Press.

Form, W. H. (1954). "Place of Social Structure in the Determination of Land Use." *Social Forces* 32: 317-323.

Goldstein, I. (1986). "The Impact of Racial Composition on the Distribution of Conventional Mortgages in the Philadelphia SMSA: A Case Study" (Working Paper). Philadelphia: Institute for Public Policy Studies, Temple University.

Gottdiener, M. and J. R. Feagin (1988). "Paradigm Shift in Urban Sociology." *Urban Affairs Quarterly* 24: 163-187.

Harvey, D. (1973). *Social Justice and the City.* Baltimore, MD: Johns Hopkins University Press.

Harvey, D. (1977). "Government Policies, Financial Institutions and Neighborhood Change in United States Cities" in M. Harloe (ed.), *Captive Cities.* London: John Wiley.

Harvey, D. (1985). *Consciousness and the Urban Experience.* Baltimore, MD: Johns Hopkins University Press.

Harvey, D. and L. Chatterjee (1974). "Absolute Rent and the Structuring of Space by Governmental and Financial Institutions." *Antipode* 6: 22-36.

Logan, J. R. and H. L. Molotch (1987). *Urban Fortunes: Political Economy of Place.* Berkeley: University of California Press.

McClory, R. (1986). *The Fall of the Fair: Communities Struggle for Fairness.* Chicago: Chicago 1992 Committee.

McKenzie, R. D. (1967). "The Ecological Approach to the Study of the Human Community" in R. Park, E. W. Burgess, and R. D. McKenzie (eds.), *The City.* Chicago: University of Chicago Press.

Molotch, H. L. (1976). "The City as a Growth Machine." *American Journal of Sociology* 82: 309-330.

Molotch, H. L. (1979). "Capital and Neighborhood in the United States: Some Conceptual Links." *Urban Affairs Quarterly* 14: 289-312.

Molotch, H. L. (1990). "Urban Deals in Comparative Perspective" in J. R. Logan and T. Swanstrom (eds.), *Beyond the City Limits: Urban Policy and Economic Restructuring in Comparative Perspective* (pp. 175-198). Philadelphia: Temple University Press.

Molotch, H. L. and J. R. Logan (1984). "Tensions in the Growth Machine: Overcoming Resistance to Value-Free Development." *Social Problems* 31: 483-499.

Molotch, H. L. and J. R. Logan (1985). "Urban Dependencies: New Forms of Use and Exchange in U.S. Cities." *Urban Affairs Quarterly* 21: 143-169.

Shlay, A. B. (1989). "Financing Community: Methods for Assessing Residential Credit Disparities, Market Barriers and Institutional Reinvestment Performance in the Metropolis." *Journal of Urban Affairs* 11: 201-223.

Shlay, A. B. (1986). *A Tale of Three Cities: The Distribution of Housing Credit From Financial Institutions in the Chicago SMSA from 1980-1983*. Chicago: Woodstock Institute.

Shlay, A. B. (1987). *Maintaining the Divided City: Residential Lending Patterns in the Baltimore SMSA*. Baltimore, MD: Maryland Alliance for Responsible Investment.

Shlay, A. B. (1988). "Not in That Neighborhood: The Effects of Population and Housing Distribution of Mortgage Finance Within the Chicago SMSA." *Social Science Research* 17: 137-163.

Shlay, A. B. (Forthcoming). "Shaping Place: Institutions and Metropolitan Development Patterns." *Journal of Urban Affairs*.

Shlay, A. B. and R. P. Giloth (1985a). "Whose World Moves the 1992 Fair?" *The Neighborhood Works* 5: 18-20.

Shlay, A. B. and R. P. Giloth (1985b). "Charting Private Interests and Public Issues: Cliques & Connections Behind the 1992 Fair." *The Neighborhood Works* 6: 10-14.

Shlay, A. B. and R. P. Giloth (1987). "The Social Organization of a Land Based Elite: The Case of the Failed Chicago 1992 Worlds Fair." *Journal of Urban Affairs* 9: 305-324.

Smith, N. (1986). "Gentrification, the Frontier, and the Restructuring of Urban Space" in N. Smith and P. William (eds.), *Gentrification of the City* (pp. 15-34). Boston: Allen and Unwin.

Stone, C. (1987). "Summing Up: Urban Regimes, Development Policy, and Political Arrangements" in C. Stone and H. Sanders (eds.), *Politics of Urban Development* (pp. 159-181). Lawrence: University Press of Kansas.

Stone, C. (1991). "The Hedgehog, the Fox and the New Politics: Rejoinder to Kevin R. Cox." *Journal of Urban Affairs* 13: 289-297.

Walker, R. (1981). "A Theory of Suburbanization: Capitalism and the Construction of Urban Space in the United States" in M. Dear and A. J. Scott (eds.), *Urbanization and Urban Planning in Capitalist Society* (pp. 383-430). New York: Methuen.

Weiss, M. A. and J. T. Metzger (1989). "Planning for Chicago: The Changing Politics of Metropolitan Growth and Neighborhood Development" in R. A. Beauregard (ed.), *Atop the Urban Hierarchy*. Totowa, NJ: Rowman and Littlefield.

Zukin, Sharon (1980). "A Decade of the New Urban Sociology." *Theory and Society* 9: 575-601.

Suggested Readings

Beauregard, R. A. (ed.) (1985). *Atop the Urban Hierarchy*. Totowa, NJ: Rowman and Littlefield.

Bradford, C. and L. Rubinowitz (1975). "The Urban-Suburban Investment Process: Consequences of Older Neighborhoods." *The Annals* 422: 77-86.

Cox, K. and A. Mair (1988). "Locality and Community in the Politics of Local Economic
Development." *Annals of the Association of American Geographers* 78 (2): 207-325.

Feagin, J. R. (1988). *Free Enterprise City: Houston in Political and Economic Perspective.*
New Brunswick, NJ: Rutgers University Press.

Giloth, R. P. and R. Mier (1989). "Spatial Change and Social Justice: Alternative Economic
Development in Chicago" in R. Beauregard (ed.), *Economic Restructuring and Political Response.* Newbury Park, CA: Sage.

Gottdiener, M. and J. Feagin (1988). "The Paradigm Shift in Urban Sociology." *Urban Affairs
Quarterly* 24: 163-187.

Harvey, D. (1973). *Social Justice and the City.* Baltimore, MD: Johns Hopkins University
Press.

Logan, J. R. and H. L. Molotch (1987). *Urban Fortunes: Political Economy of Place.*
Berkeley: University of California Press.

Molotch, H. L. (1976). "The City as a Growth Machine." *American Journal of Sociology* 82:
309-330.

Shlay, A. B. (1985). *Where the Money Flows: Lending Patterns in the Washington, D.C.-
Maryland-Virginia SMSA.* Chicago: Woodstock Institute.

Shlay, A. B. (Forthcoming). "Shaping Place: Institutions and Metropolitan Development
Patterns." *Journal of Urban Affairs.*

Shlay, A. B. and R. P. Giloth (1987). "The Social Organization of a Land Based Elite: The
Case of the Failed Chicago 1992 Worlds Fair." *Journal of Urban Affairs* 9: 305-324.

Smith, M. P. and J. R. Feagin (eds.) (1987). *The Capitalist City: Global Restructuring and
Community Politics.* Oxford: Basil Blackwell.

Stone, C. and H. Sanders (eds.) (1987). *The Politics of Urban Development.* Lawrence:
University Press of Kansas.

Zukin, Sharon (1980). "A Decade of the New Urban Sociology." *Theory and Society* 9:
575-601.

9

Race and Class in Local Economic Development

JOHN J. BETANCUR
DOUGLAS C. GILLS

The Role of Theory in the Local Economic Development Efforts of Racial "Minorities"

Locally based development efforts of national or racial "minorities"[1] in U.S. cities have not been so inspired by "academic theories" as they have been the result and basis of social praxis. In other words, the efforts are driven more by practical circumstance—struggle and contention—than by abstract theory.[2] So-called minority development largely consists of efforts to obtain freedom, justice, or equality for or among people who have lived the reality of being Latino, black, or Native American in a society dominated by whites in power. Such efforts, in fact, have been forced by objective realities of exclusion and domination based on race and nationality.

Two contradictory forces have shaped the outcome of "minority" development efforts. First is the demand for effective equality by these oppressed groups and their socially conscious advocates. Second are the accommodating responses of those in positions of power, wealth, and privilege. These forces have met somewhere between the two extremes (Stannard, 1992).

The "majority" often have yielded, through public policy and incremental and limited concessions. While accepting the need for special efforts, they have advanced programs that generated small-scale, marginal changes while favoring individual access. Their underlying rationales have led to stratagems of social control or integration/assimilation "on the white man's terms."

In contrast, the oppressed have promoted *group* strategies and have been effective in mobilizing on the basis of self-help, protest, and collective politics. They have struggled to achieve and use power for advancement of their *group* agendas, have used physical concentration for some economic

191

or business development, and have tried to control and reform local institutions to facilitate collective development in their communities. In short, they have used all resources at hand, including their size, to advance their interests.

Operating within a highly individualist society and economy, oppressed nationals often have been "competitively divided" by the dominant majority. Many of them are co-opted into pursuing individual gains at the expense of the community. While collective action opened doors for them to enter into mainstream institutions or obtain public and other benefits, their successes actually became a substitute for the group's demands. Indeed, many of these individuals ideologically identify with their new class cohorts while reaping the benefits of their "minority" status in the mainstream.

As a result, these oppressed groups have been polarized into a small aggregate of higher income earning individuals and a large mass of poor people being increasingly marginalized. The former, as Wilson (1987) observes, have moved into the middle- or upper-income ranks and into exclusive neighborhoods. The latter have been further "ghettoized" in the inner city and into some suburbs characterized by disinvestment. Such an outcome can be largely explained by the nationality or race framework. Indeed, the struggle has been predominantly focused on racial or nationality oppression to the neglect of the overlapping class oppression that further complicates their condition.

√This analysis acknowledges that class forces are a central dynamic in urban economies and policy interventions. The tremendous racialization of U.S. development policy, however, has overshadowed class. The often exclusive prevalence of race in urban planning and development politics is alarming and disturbing. It, in fact, calls for new critical thought, summation, and approaches that push the policy struggle beyond the limitations of prevailing neighborhood-, race-, or nationality-based efforts. While the majority is still largely in control, progressive coalitions—such as the one underpinning the administration of the late Mayor Washington in Chicago—illustrate the strength of all the oppressed—class, race, nationality, and gender groups—when acting as groups in collaboration. The most recent trauma[3] in Los Angeles also shows how devastating ghetto-barrio eruptions can be—whether viewed as legitimate or misdirected. As long as the system of exclusion, segregation, and labor segmentation is kept in place, social tension will continue and will likely produce renewed eruptions, even if many are likely to be shortsighted. So long as the struggle excludes the class dimension and is dominated by the search for upward mobility, co-optation and accommodation will continue weakening it, in fact, will further push it away from the class struggle.

We, furthermore, recognize the prevalent assertion that the gains of oppressed groups are often perceived as threats to the dominant majority

(Hacker, 1992). As a result, urban development activity and planning assume the form of a struggle over whether the privileges of the majority or the demands of the oppressed will prevail. These contentions are played out within the urban centers, where the latter are most concentrated. The issues discussed next reflect this tension.

Four main dichotomies—between individual and group-based development, concentration and dispersal, nationality and majority control, and growth and equity—are at stake here and set the stage for examination of "minority" models of local economic development.[4]

Polarities of "Minority" Development

Nationality Group-Based Versus Generic/Individual Development

A first dispute underlying nationality-based development analyses is that of the relevance of a "minority" group focus. On the one hand, there is the prevalent belief that there should be no *group* claims to special needs and assistance. The system should reward individual hard work, industriousness, and thrift regardless of nationality. Communities should be able to prosper on the basis of the hard work of their individual members. With the enactment and implementation of civil rights legislation, individuals, not groups, should be able to rise according to merit. Recent Supreme Court rulings under the Reagan-Bush regimes have supported this "conservative" view.

This belief ignores that blacks were enslaved *as a group* and treated legally *as a group*. Mexicans and Puerto Ricans were imported for certain types of jobs and turned into low-wage laborers in segmented markets, worked under unique conditions of servitude, and have been excluded *as groups*. Native Americans, in turn, were massacred, expropriated, and systematically oppressed *as groups*. This condition has been historically maintained to the point that today these groups rank among the poorest *groups* in the nation and are relegated to living on America's versions of "homelands."

Standing astride this conflict over group status is an emerging "neoliberal" view, which argues that neighborhoods, families, and social networks should be added to nationality as factors for explaining the situation of "minorities" (Wilson, 1991). Concentration of the unskilled, the uneducated, and the un/underemployed in ghettos along with female-headed households and social outcasts produces mutually reenforcing patterns of behavior and poverty. As a result of this, ghettos/barrios become material and psychological traps for their residents.

This view tends to minimize the issue of *group* oppression while supporting the promotion of "nonsolutions" such as "role models" and "dispersion." Indeed, it has provided new ammunition for those who replace oppression with generic categories of "poverty" and "poverty culture," "homelessness," "human capital," "family disintegration," and other "social ills." Some even argue that a "minority" status is a handicap or results in self-fulfilling prophesies, because it may lead to self-segregation or self-pity. Indeed, it may become an excuse for *not trying hard enough* or for adherence to values or habits *opposed to those needed for success*, as is sometimes argued (Sowell, 1975).

Community-Based Development Versus Geographic Dispersal

U.S. cities are highly segregated by nationality. Blacks and Latinos have become particularly clustered into single-race, identifiable geopolitical areas defined by nationality (Goldsmith and Blakely, 1992). Such areas are typically disinvested in or redlined to the point of near devastation. Too frequently, development bypasses them or unregulated reinvestment displaces their indigenous people, particularly those on the margins of economic life in the cities.

Some view the concentration of oppressed groups as a unique opportunity for locally based development focused on the specific needs and conditions of such groups. Relatedly, groups have often used community-based development initiatives to organize and struggle as a group, gain a voice, develop institutions, establish greater group representation, and gain greater control of resources through collective local development projects.

Others, following both the individualistic and the neoliberal traditions, argue that the "inner city" lacks the capacity to absorb investment, support businesses, or maintain improved housing and facilities (Crandall and MacRae, 1971; Brimmer and Terrel, 1971). Dispersal, on the contrary, they add, helps break up poverty cycles, eliminates ghettos, exposes "minorities" to opportunities, and provides success-oriented role models. A now classical example of this view is the federal court-directed program in Chicago that relocates public housing residents into nonghetto areas as an effort to deconcentrate poor "minority" families by encouraging their relocation to "general areas" of the city and suburbs.

Nationality- Versus Majority-Controlled Development

The nationality model parallels the community control model. Frustrated by paternalism and other demeaning initiatives of the white majority, oppressed groups responded by acting on the need to control their own process of development or to assume positions of power and authority.[5] Not only do externally controlled initiatives ignore their specific conditions but their proponents lack the required sense of urgency and group development of oppressed nationality groups. In Chicago, the (white) democratic political machine extended participation to some individual African Americans and Latinos on its terms, while neglecting the needs of their respective constituencies (tokenism). This is a clear example of development "on the white man's terms." The "Council Wars" in the same city (1983-1986) also provide a dramatic illustration of efforts to break up this historic practice by the reform government of Mayor Washington (Alkalimat et al., 1984).

For many black and Latino activists, community control parallels the movement for community-directed development now prominent in local development literature. Control of development is particularly relevant in group-specific initiatives. It is a call for self-determination, equitable distribution of opportunities and benefits, and democratization of decision making. This orientation is prevalent among activists in diverse settings and neighborhoods. For the oppressed, however, it has a particular importance because development has at the same time a nationality character. Class oppression in America is often rationalized in nationality, racial, or ethnic terms.

Oppressed groups' claims for control are often viewed with suspicion by the majority group representatives, who tend to resent public policy efforts to adjust for institutionalized exclusion. In fact, the prevailing sentiment is to dismiss their demands as those of "special interests."

In sum, the argument is that oppressed groups know best their conditions and needs and can best represent their own interests, while dominant group actors are self-serving and their works tend to reproduce the status quo.

Growth Versus Equity and the New Populism

Most local development approaches have been dominated by a "growth mentality" (Squires et al., 1987) under the claim that development/growth is a strategy with no losers because its benefits "trickle down" to everybody. Evidence, however, shows that recent growth in cities like New York or Los Angeles did not improve the conditions of oppressed groups (Sassen, 1990). On the contrary, it increased polarization and was accompanied by losses in the comparative position of these groups as well as in the resources going to their communities. A wealth of evidence has been gathered also that contradicts the "trickle down" claim. Under the current conditions, growth makes developers more wealthy, while making the poor, and oppressed, groups poorer (Harrison and Bluestone, 1990; Squires et al., 1987).

Given their systematic exclusion and economic discrimination, oppressed groups have placed emphasis on the need for equity as a development goal and strategy. Equity enhancement has not been a priority for mainstream, progrowth local economic development platforms (Squires et al., 1987). The populist agenda stresses equity before growth. Why should the people pay or sacrifice to promote private development initiatives with government support when many of these projects feasibly can be developed without such support and when oppressed communities derive such limited benefits and access from the projects (Gills, 1991)?

Prevailing Local Economic Development Models

The four tensions we have discussed are present in the three prevailing models of development emanating from oppressed communities.[6] One is defined by a strong group identity and has a "community" focus and a quest for social equity. A second is defined by an individualistic orientation, an emphasis on "community" as a vehicle for mainstreaming with an ultimate goal of dispersal, and a reliance on growth. Juxtaposed against both of these is an orientation defined by nationality and community control of development. Each of these constellations of emphases constitutes a model of development. Each model contains some element of "community," yet the models mean very different things. Following is a discussion of the models of *community development, affirmative action,* and *nationalism/Black capitalism.*

Community Development

The call for community economic development dates back to the 1960s. It was largely a response to the riots and rebellion in the central-city ghettos of that period. The Special Impact Program of the Office of Economic Opportunity (OEO) was the first and the largest effort. Skeptical about the possibility of a quick, effective social and economic integration of blacks, the National Advisory Commission on Civil Disorders called for large-scale improvements in the quality of ghetto life as the best option to quell social discontent (1968).

Indigenous institutions were proposed by some as the most effective mechanism for bringing about this end. They would be the basis for developing viable local economies and human capital, while building the necessary political base and empowering residents (Vietorisz and Harrison, 1971). In community after community, and in city after city, rioting and disorder were met with "community development" packages.

The community development approach is not without internal tensions. From the perspective of society-at-large, when not linked to a strategy of liberation, it can operate as a form of *social control*. From the perspective of oppressed groups, it has been a platform of potential struggle and mobility. It therefore is not a single, unified strategy. Stratagems and models range from mainstream to social-change orientations.

Community development has been presented as a vehicle for revitalizing low-income communities within a framework of corrective capitalism. The community development corporation (CDC) has emerged as the main instrument of this undertaking (Peirce and Steinbach, 1987). In cities like Chicago, CDCs have been very prominent in "minority" areas and have engaged in a wide range of activities including real estate, employment, and business development, land trust and investment initiatives, and productive enterprises (Betancur et al., 1991).

In addition to being an alternative mechanism for ghetto development or self-sufficiency enhancement, it is also pursued by many as a democratic, populist, pro-neighborhood alternative to big-time government and corporations. Indeed, it links development objectives with political empowerment goals. It represents a bloc of interests of poor and working-class residents outside the sphere of production, fighting against urban domestic expenditure retrenchment, for a more equitable distribution of public resource allocations, and for an increasing share of fiscal expenditures to be directed toward neighborhoods—*under conditions of local control*. These ideas are particularly appealing to members of oppressed groups who have been routinely excluded from making decisions concerning their welfare and their communities and who view local control as a stepping stone for extending democratic practices.

From a more extreme perspective, some "minority" leaders and others have advocated using community development as a transitional approach toward a socialist alternative. They claim that capitalism keeps oppressed groups at a competitive disadvantage and "competitively divided" (Hampden-Turner, 1969), making things worse for them. Thus they propose a range of initiatives within which communities are the basis for change and the units for development of new relations leading to the construction of a new society.

The community development focus has been enriched by analyses such as the "internal colony," the "drain of resources," and the "ghetto" metaphors. The "internal colony" discussion has brought attention to the domination of nonwhites by whites in urban America; the forceful segregation of oppressed groups; outside control of local institutions by the (white) majority; and the use of these institutions to legitimize economic and social inequalities, while reproducing the relations of domination (through school, police, welfare, and other systems).

The "drain of resources" metaphor shows the absence of local or intragroup multiplier effects and the leakage of local or group resources through a series of mechanisms, ranging from disinvestment and redlining to business and factory flight.

Finally, the term *ghetto* emphasizes the isolation and exclusion of oppressed groups from the city and its resources and from society as a whole as well as the physical confinement of its inhabitants, who have minimal options to leave and to transform it into a viable space for people to live and work (Gunn and Gunn, 1991).

Affirmative Action

Affirmative action has emerged as one of the strongest integrationist models. Oppressed groups have consistently included this element in their struggles, in an effort to gain access to jobs and positions, particularly in government and in firms with a high sensitivity to boycotts. Affirmative action compliance has led to the relative expansion of nationality middle classes as well as the proliferation of firms receiving public contracts, set-asides, and distributorship, including women-led firms.

Many employers' response to it has been one of tokenism, labor segmentation, and adherence to the minimum standards. Others have argued that it creates an undue burden as it is often hard for them to identify "qualified" members of these groups in pursuit of quantified objectives. It is also blamed for problems of sinking morale among (white) majority employees or is viewed as "reverse discrimination." Finally, there are those who oppose nationality as a factor in the workplace and claim that people should gain position and move up the ladder according to merit and hard work, not according to nationality,[7] as noted earlier. The assumption is that nationality oppression is a relic of history rather than a life force in current public policy.

Separatism

In contrast to the prevailing integrationist model advocating for the incorporation of oppressed groups *as individuals* in the opportunities and benefits of society at large, separatism calls for an autonomous effort to develop power, wealth, and human capital among such *groups*. It also stands apart from development efforts that do not incorporate strong demands for community control.

Perhaps the best known model is "Black Capitalism." This idea has inspired a range of efforts from a request for a full-fledged program of government support for black business development, through "buy black" type efforts, to autarkic black development. This idea had its radical counterpart in proposals for "socialism without nationalism" based on black separation, worker control, economic planning, and cooperation (Boggs, 1971). Although Black Capitalism is a separatist model, there is an integrationist variant that is tied to entrepreneurialism and middle-class elitism (Ofari, 1970; Allen, 1970).

The logic underpinning this model is equally applicable to Latinos, Native Americans, and others. In fact, these groups have also supported strategies facilitating business development and wealth creation in the hands of their members (or some strata among them). Neighborhood economic enclaves (retail in particular) have helped them and other individuals to accumulate some wealth and have been the basis for expanding into other markets and eventually becoming entrepreneurs on their own.[8] The down side is that this stratum does not view itself as accountable to the respective community.

"Nationalism" is an orientation calling for strategic efforts by oppressed groups to gain political or economic power and use it to promote their causes. It is based on the assumption that these groups are excluded as *groups*, not as individuals. Only if they achieve power as groups will they be viewed with respect and will they be able to confront the white majority (Jennings, 1992). This element has been extremely important in the struggle for drawing the boundaries of political units (wards, congressional districts) to coincide with the areas of concentration of oppressed groups. The purpose is to have members of these groups represent their people and to eventually gain control of representative bodies, especially at the local level. The fight for "community control" by local residents and the current thrust toward *community-directed economic development* manifests this orientation.

Notice that we differentiate between nationalism (ideological) as a worldview or political philosophy and nationality or national development as a historical-social category. We also distinguish between nationalism of the oppressed and the oppressor (Alkalimat et al., 1984). Hence, when we talk about *nationality groups,* we refer to those people that are oppressed on the basis of their national origin and historical-cultural backgrounds. When we discuss nationalism as a strategy, we are addressing the struggle of a people to gain political power *as a group* and use it in pursuance of its collective interests. Finally, there is oppressive nationalism or the exercise of European ethnic monopoly of power to the exclusive benefit of its group members and the suppression of others. Clear examples of this are local political machines, unions, and organized crime syndicates rooted in ethnically-based gangs.

Local Development Theory in Practice:
The 1992 World's Fair Development Project and
Minority Perspectives

This section illustrates the previous "theoretical" elements in practice through their application to the proposed 1992 Chicago World's Fair. We will discuss this around the three main models of "minority" economic development at the local level, namely, community development, affirmative action, and separatism.

Background

The proposed 1992 world's fair struggle dominated the development agenda in Chicago during the early to mid-1980s. The debate around the fair was a classic example of the contention between growth coalitions evident in many cities and various contending, but distinct, community development and "minority" interests. The story of the fair—as an early 1980s vision of the growth coalition supported by former Mayor Jane Byrne and the subsequent struggle with the progressive administration of Mayor Washington—has been told elsewhere (Shlay and Giloth, 1987; Mier, forthcoming). What is important here is the interplay of the fair planning interests and their opposition from the perspective of the models we have been discussing.

The fair was considered a secondary, even tangential issue relative to the election of Mayor Washington. It was discussed, and after his election, a committee and a working policy were developed to deal with it. Washington stayed neutral for a long time and, finally, indicated that he endorsed the fair "with verve, vigor, and gusto" (McClory, 1986: 22). He stated that so long as the fair was able to pay its own bills and did not impede the city from a fair distribution of services and resources to the neighborhoods, he would not oppose it. This was consistent with his policy advocacy of balanced growth between the central business area and the needs of the neighborhoods.

The Fair and Community Development

At the time of the fair proposal, there were growing and energetic nonprofit community development groups that had by 1980 built interlocking coalitions and trade associations. The core activists within these overlapping networks were drawn from 125 to 150 community-based organizations serving a wide range of constituents and organized interests concerned with community development in the neighborhoods as a priority of local public policy. They were supported by technical assistance providers to the community-based development movement. These diverse organizations formed an ad hoc coalition called the "Chicago 1992 Committee," which rallied under the cry for "fair, Fair" policy.

The committee was largely the result of community-based organizing efforts initially launched in "minority" communities such as Pilsen with a tradition of struggle against the repeated efforts of downtown interests to turn surrounding areas into New Towns in Town.[9] It was first housed at the Latino Institute and had a strong Latino and black presence. The determination and concerns of black and Latino communities most directly affected by the fair to fight displacement gave the coalition particular energy and urgency.

This rainbow movement organized to oppose or to change Mayor Byrne's developer-centered policy thrust and subsequently supported Washington's successful challenge to Byrne in 1983. What is impressive about this network, with authentic grass-roots bases and linkages, is that it represented the most promising emergence of a multinational, interracial, low-income, *community-based* policy front to date (Clavel and Wiewel, 1991).

The minimum demand of the majority of the coalition membership was for a true picture of the fair's costs and who would bear them. A subordinate position, held by a minority, was that the fair should be opposed if it would deplete developmental, infrastructural, and public service resources to the neighborhoods; that it could be supported to the extent that it did not adversely affect them; and that its residuals were shared in proportion to the needs of and adverse effects on areas around the site. A third position, held by a radical minority, was that the fair promoters were not credible and—on these grounds alone—it should be opposed. When the fair was eventually killed in 1985, after repeated evidence of escalating costs and diminishing impact, most community-based interests, including the Chicago 1992 Committee, thought it a victory for community development forces and their constituents.

Black and Latino communities adjacent to the site were particularly thrilled because they represented the forefront of local activism and had strongly advocated for a development model based on community and opposed to displacement/gentrification.

The Fair and Affirmative Action

Prominent black and Latino leaders and civic or community activists (particularly from the business sector) broke with the community development viewpoint and politics with respect to the fair. Groups such as the Urban League, Operation PUSH, and the Latino Institute argued for a pragmatic position of ensuring adequate "minority" participation in the enterprise. They were convinced that the fair would take place against any and all opposition. They then demanded an affirmative action agreement/covenant and asked for independent groups to monitor compliance.

It would be unfair to say that there were no overall concerns about the impacts, opportunity costs, and implications of the fair upon their broader "minority" communities. Assuming again the imminence of the fair, however, these organizations and individuals argued that Mayor Washington should use his leverage to broker concessions from the World's Fair Authority and its developers and promoters that would meet the needs of "qualified" (meaning professional) "minorities" as managers, workers, contractors, vendors and suppliers, exhibitors, and entertainers.

The Fair and Separatism

The dominant interests of the black nationalist view on the world's fair were reflected through the "Task Force for Black Political Empowerment" (or "task force") and the Chicago Black United Communities (CBUC). Among Latinos, the Mayor's Advisory Committee on Latino Affairs (MACLA) was perhaps the best organized and strongest expression of Latino "nationalism."

The predominant view among nationalists within the task force[10] was that Washington should oppose the fair on the grounds that (a) it was Byrne's and a white developers' project from which blacks were largely excluded and (b) it was another scheme to dislodge black people from the southern lakefront.

Between 1983 and 1984, there was a slight shift in this position. It seemed increasingly likely that the fair was going to happen. Its promoters were furiously courting black—and, to a lesser extent, Latino—leaders and seeking their input on affirmative action and "minority" participation. The results were a split in points of view: a waning of active public responses and an increased involvement of task force members in meetings, hearings, and conferences sponsored by the World's Fair Authority. Partly because of uncertainly about the administration's intentions and partly to minimize a division in the task force, it agreed to apply the recently developed "Black Agenda for Political Empowerment" as a screen to view community participation of blacks in job opportunities, contracts and concessions, and exhibition space (to promote the group's culture and history within the sites). Finally, the task force would work to assure black women participation and to minimize disruption/destruction of the black residential community.

From all indications, the World's Fair Authority was amenable to negotiation on these matters, provided that there was a fair and that the task force representatives were supportive of the fair taking place. The task force, formally, made no deal on this latter point, asserting, tactfully, that it could take no public position that would embarrass Mayor Washington.

While the majority within the task force supported political expediency, others within and outside the task force adhered to the original formulation. Ideologically, some members within the task force and CBUC opposed the fair, in principle. This minority view asserted that the 1992 celebration of Western colonialism was an insult to black people and Native Americans. It is noteworthy that the proponents of this position made few efforts to network with Chicago's large, if dispersed, Native American community.

Meanwhile, the threat of displacement of a Latino stronghold, the Pilsen neighborhood, turned the fair into an immediate target of Latino "nationalist" interests. Since the mid-1970s, the local leadership had expressed its determination to keep Pilsen Latino and to oppose any attempts at displacement (Mier and Moe, 1991). Organized around MACLA,[11] Latinos held a hearing in Pilsen in September 1984. The result was almost unanimous opposition to the fair. The opposition was based on the determination to keep Pilsen Latino, the experience of being ripped off by downtown interests, and the call for local self-determination of the Mexican community. While groups such as Pilsen Neighbors tried to get a piece of the action—in the form of jobs, contracts, and other commitments—more Latinos felt that the fair was not the kind of development beneficial to the community. Indeed, it was owned and controlled by majority white interests who would expropriate most of the benefits at the expense of Latino and black concerns.

The Fair and Models of "Minority" Development

What we have gleaned from the reexamination of this recent development policy debate are the following points as they relate to local "minority" development models in practice and their linkage to larger struggles.

First, the force underlying the fair confrontation was the contradiction between growth coalitions and pro-neighborhood forces over the public economy. While growth coalitions were pressing the state to assume the risks and subsidize their highly speculative activities, pro-neighborhood forces were advocating for the use of public resources to develop viable neighborhoods.

Second, decades of neglect brought a multiplicity of neighborhood interests together to oppose still one more mega-project threatening more than a few nationality communities. The communities most immediately affected by the fair played a central, energizing role in mobilizing opposition to the fair developers. This coalition, moreover, was multinational and multiracial and included multiple interests. Groups were united in their opposition to growth coalition politics and in their interest in neighborhood development—as vague as this formulation was. The community coalition was not united by a theory bringing together class struggle, neighborhood, and nationality or racial interests or by a clear development alternative building bridges across racial boundaries.

Third, many interests were presented as nationality or racial group interests and, indeed, concealed class issues, particularly through nationality or locality formulations. The central role of "eliticized" individuals profiting from concessions, such as government contracts, often led to pragmatic positions of compromise at the expense of the interests of grass-roots community interests.

Finally, the form of participation of nationality groups revealed once more the absence of a coherent strategy and theory of community-based analysis and development. It illustrated the disjuncture between nationality and class struggles and the absence of a solid theory linking race/nationality and class. Divisions within "minority" groups manifested contradictory interests between co-opted subgroups and the broader constituency bases. The assumption that it is better to get something, anything, rather than to be left out altogether is clearly a pragmatic position inspired by the belief in trickle down economics. It in fact demonstrates the lack of a solid class- and race-/nationality-based position. Such practices do not successfully challenge the legitimacy of race and class domination.

Theories of Race and Class in Local Development

Recent economic decline has seen the vicious return of racism and the erosion of previous civil rights gains (Johnson et al., 1992). This situation further challenges the effectiveness of conventional policy models and other safety-net programs of improvement for oppressed nationalities. Some coalition movement initiatives have showed more potential than others for collective or individual gain. Perhaps the most prevalent, still important, and most underfunded and unrecognized are community-based development initiatives. Communities have been the main support bases for the empowerment and struggle of oppressed groups in recent history in the United States. Strong political movements among these groups have started there. Moreover, political movements have maintained their vitality on the basis of grass-roots organization. In Chicago, community-based black and Latino politicians were able to coalesce with other reform forces to challenge the democratic machine and were close to forming a new community-based form of local governance (Clavel and Wiewel, 1991). Through political control, these efforts were able to produce job and business opportunities for members of their oppressed communities. Moreover, with government support they could attract important investment to the inner city. Most of this work was locally based and relied largely on the strength of the communities represented and their organization rather than on their economic resources.

Apart from the realities of the uneven development of capitalism and the mechanisms reproducing the overall role of oppressed groups as cheap labor, four main forces further undermine the potential of this approach. First, investment in these communities may be a waste if it is not accompanied by *collective* improvements in the generalized economic conditions of the vast majority of its residents. Second, the current pervasive dependency of CDCs on outside support, particularly on government sources, makes them extremely sensitive to the will of the political and foundation community to continue supporting them. Third, as CDCs increase their scale and effectiveness, they are likely to be scrutinized more closely. Today, they remain the targets of local growth machines and of opposition by mainstream local economic actors who charge them of "unfair competition" or of being "antidevelopment" or who demand a share of the business (via privatization, public-private partnerships, and so on). Finally, CDC activists have not advanced a consistent theory and strategy of development that breaks with conservatism on the one hand and patronizing liberalism on the other.

Progressive-minded people of nationalities and populations must be active in this fight for it is rooted in self-enlightened interest. Even though racial domination and class domination are not the same, and the elimination of one will not automatically lead to the elimination of the other, their convergence creates the opportunity to tie them together and build struggles for a more comprehensive goal—the end to forms of social domination, oppression, and exploitation and the building of viable communities with enhanced social development of human capacities.

Notes

1. Although we use the term *minority groups,* we prefer the terms *nationality* and *oppressed groups.* These terms better express the conditions of oppression along national lines in the United States.

2. There is pervasive ambiguity in what is termed *theory.* What we want to emphasize here is that, different than development of theories within a predominantly academic environment, most of the thinking about local economic development related to minorities has evolved from the struggle itself within the actual relationships of domination discussed in this chapter.

3. African American and Hispanic residents of Los Angeles refer to it as a "rebellion" (Johnson et al., 1992).

4. Although class dynamics are fundamental to development politics, we focus upon the "nationality" category. We do so recognizing that there are real, critically important internal class tensions and divisions within nationality communities.

5. For recent works on nationality empowerment, see Villareal and Hernández (1991), Jennings (1992), and Marable (1985).

6. We use the term *model* more as an expression of tendencies than as discrete categories. The boundaries between the "models" we propose are fuzzy, and, in any instance, they can closely reflect each other.

7. This claim by the dominant majority becomes a code for further exclusion. In fact, historic oppression has prevented "minorities" from developing competitive capacity and has built an accumulated advantage for the majority. In this way, when race or nationality are discounted as factors of exclusion, the white majority continues to retain the competitive advantage.

8. The black capitalist model is not new, however. Its origins go back to the early twentieth century when blacks took advantage of segregated markets to which they were restricted as business owners. A leading activist advocate was Booker Washington.

9. Pilsen residents had experienced displacement with the construction of the University of Illinois campus. Threatened by the Chicago 21 Plan, they had organized the oppositional Pilsen Planning Coalition. Organized again against the fair, the Pilsen Planning Coalition got together with organizations in other affected communities, especially black communities to the south of the fair's site. They then decided that a Latino-black coalition could not possibly have the impact that was necessary to successfully oppose the fair. At this point, they started contacting community networks, individuals, and organizations throughout the city. The process profited from an emerging community movement and ended with the formation of the Chicago 1992 Committee.

10. TBPE was organized as the informal arm of the Washington campaign in November 1982 (Gills, 1991). It had been the most significant, representative, and effective of the movement groups that promoted and actively supported Washington's candidacy and election. While it was broad based in leadership representation, it was dominated by political reformers and community-based activists in the early stages of the campaign and by the African American nationalist leadership subsequent to Washington's election.

11. Largely excluded during the administration of former Mayor Richard J. Daley, Latinos have engaged in independent political efforts since the 1970s. In 1981 blacks and Latinos came together in a historic effort to challenge state and city redistricting, which culminated with special elections in 1986. These and other initiatives of affirmative action, defense against developers, and education stirred the community and gave a unique organizational impetus to the group. Encouraged by the candidacy of Harold Washington and his success in the primaries, the bulk of the Latino leadership supported him and called for fairness in the distribution of municipal services, employment, and other opportunities. The election of Washington consolidated these efforts and led to the formation of a "Latino Platform." At the core of this platform and efforts were affirmative action in contracts, hiring, and appointments; neighborhood-oriented economic development; and Latino empowerment.

References

Alkalimat, Abdul & Associates (1984). *Introduction to Afro-American Studies*. Chicago: Peoples College Press/21st Century Books.

Allen, Robert (1970). *Black Awakening in Capitalist America*. Garden City, NY: Doubleday/Anchor.

Betancur, John J., Deborah E. Bennett, and Patricia A. Wright (1991). "Effective Community Strategies for Community Economic Development" in Phillip W. Nyden and Wim Wiewel (eds.), *Challenging Uneven Development: An Urban Agenda for the 1990s*. New Brunswick, NJ: Rutgers University Press.

Boggs, J. (1971). "The Myth and Irrationality of Black Capitalism" in R. E. Bailey (ed.), *Black Business Enterprise*. New York: Basic Books.

Brimmer, A. F. and H. S. Terrel (1971). "The Economic Potential of Black Capitalism." *Public Policy* 19 (Spring): 289-308.

Browning, Rufus, Dale R. Marshall, and David Tabb (eds.) (1990). *Racial Politics in American Cities*. New York: Longman.

Clavel, Pierre and Wim Wiewel (eds.) (1991). *Harold Washington and the Neighborhoods: Progressive City Government in Chicago: 1983-1987*. New Brunswick, NJ: Rutgers University Press.

Crandall, R. and C. D. MacRae (1971). "Economic Subsidies in the Urban Ghetto." *Social Sciences Quarterly* 52 (December): 492-507.

Gills, Doug (1991). "Chicago Politics and Community Development: A Social Movement Perspective" in Pierre Clavel and Wim Wiewel (eds.), *Harold Washington and the Neighborhoods*. New Brunswick, NJ: Rutgers University Press.

Goldsmith, William and Edward Blakely (1992). *Separate Societies: Poverty and Inequality in U.S. Cities*. Philadelphia: Temple University Press.

Gunn, Christopher and Hazel Dayton Gunn (1991). *Reclaiming Capital: Democratic Initiatives and Community Development*. Ithaca, NY: Cornell University Press.

Hacker, Andrew (1992). *Two Nations: Black and White, Separate, Hostile, Unequal*. New York: Scribner.

Hampden-Turner, Charles (1969). "Black Power: A Blue-Print for Psycho-Social Development" in R. Rosenbloom (ed.), *Social Innovation in the City*. Cambridge, MA: Harvard University Press.

Harrison, Bennett and Barry Bluestone (1990). *The Great U-Turn*. New York: Basic Books.

Jennings, James (1992). *The Politics of Black Empowerment: The Transformation of Black Activism in Urban America*. Detroit, MI: Wayne State University Press.

Johnson, James H., Cloyzelle K. Jones, Walter C. Farrell, Jr., and Melvin L. Oliver (1992). "The Los Angeles Rebellion: A Retrospective View." *Economic Development Quarterly* 6 (4): 356-372.

Marable, Manning (1985). *Black American Politics*. London: Verso/New Left.

McClory, Robert (1986). *The Fall of the Fair: Community Struggle for Fairness*. Chicago: Chicago 1992 Committee.

Mier, Robert (Forthcoming). "Economic Development and Infrastructure" in David Perry (ed.), *Building the Public City*. Newbury Park, CA: Sage.

Mier, Robert and Kari J. Moe (1991). "Decentralized Development: From Theory to Practice" in Pierre Clavel and Wim Wiewel (eds.), *Harold Washington and the Neighborhoods* (pp. 64-99). New Brunswick, NJ: Rutgers University Press.

National Advisory Commission on Civil Disorders (1968). *Riot Report*. New York: Bantam.

Ofari, Earl (1970). *The Myth of Black Capitalism*. New York: Monthly Review Press.

Peirce, Neal and Carol Steinbach (1987). *Corrective Capitalism*. New York: Ford Foundation.

Shlay, Anne and Robert Giloth (1987). "The Social Organization of a Land-Based Elite: The Case of the Failed Chicago's 1992 World's Fair." *Journal of Urban Affairs* 9 (4): 305-324.

Sowell, Thomas (1975). *Race and Economics*. New York: David McKay.

Squires, Gregory, Larry Bennett, Kathleen McCourt, and Phillip Nyden (1987). *Chicago: Race, Class, and the Response to Urban Decline.* Philadelphia: Temple University Press.

Stannard, David E. (1992). *American Holocaust: Columbus and the Conquest of the New World.* New York: Oxford University Press.

Vietorisz, T. and B. Harrison (1971). "Ghetto Development, Community Corporations, and Public Policy." *Review of Black Political Economics* (Fall): 28-43.

Villareal, Roberto E. and Norma Hernández (eds.) (1991). *Latinos and Political Coalitions.* New York: Praeger.

Wilson, William Julius (1987). *The Truly Disadvantaged.* Chicago: University of Chicago Press.

Wilson, William Julius (1991). "Poverty, Joblessness, and Family Structure in the Inner City: A Comparative Perspective." Unpublished paper.

Suggested Readings

Abraham, Kinfe (1991). *Politics of Black Nationalism: From Harlem to Soweto.* Trenton, NJ: Africa World Press.

Balibar, Etienne and Immanuel Wallerstein (1991). *Race, Nation, Class: Ambiguous Identities.* New York: New Left.

Ball, Wendy and John Solomus (eds.) (1990). *Race and Local Politics.* London: Macmillan.

Boggs, James (1970). *Racism and the Class Struggle.* New York: Monthly Review Press.

Browning, Rufus, Dale R. Marshall, and David Tabb (eds.) (1984). *Protest Is Not Enough.* Los Angeles: University of California Press.

Clavel, Pierre and Wim Wiewel (eds.) (1991). *Harold Washington and the Neighborhoods: Progressive City Government in Chicago: 1983-1987.* New Brunswick, NJ: Rutgers University Press.

Davis, John E. (1991). *Contested Ground: Collective Action and the Urban Neighborhood.* Ithaca, NY: Cornell University Press.

Franklin, Raymond S. (1991). *Shadows of Race and Class.* Minneapolis: University of Minnesota Press.

Giloth, Robert and Robert Mier (1989). "Spatial Change and Social Justice: Alternative Economic Development in Chicago" in Robert Beauregard (ed.), *Economic Restructuring and Political Response.* Newbury Park, CA: Sage.

Goldsmith, William and Edward Blakely (1992). *Separate Societies: Poverty and Inequality in U.S. Cities.* Philadelphia: Temple University Press.

Hero, Rodney (1992). *Latinos and the U.S. Political System.* Philadelphia: Temple University Press.

Hogan, Lloyd (1984). *Principles of Black Political Economy.* Boston: Routledge & Kegan Paul.

Jencks, Christopher (1992). *Rethinking Social Policy: Race, Poverty, and the Underclass.* Cambridge, MA: Harvard University Press.

Katznelson, Ira (1992). *Marxism and the City.* Oxford: Clarendon.

Marable, Manning (1983). *How Capitalism Underdeveloped Black America.* Boston: South End.

Marable, Manning (1992). *The Crisis of Color and Democracy.* Monroe, ME: Common Courage.

Murray, Charles (1984). *Losing Ground*. New York: Basic Books.

Orfield, Gary and Carole Ashkinaze (1991). *The Closing Door: Conservative Policy and Black Opportunity*. Chicago: University of Chicago Press.

Pinkney, Alphonso (1992). *The Myth of Black Progress*. New York: Cambridge University Press.

Pohlman, Marcus (1990). *Black Politics in Conservative America*. New York: Longman.

Preston, Michael, Lenneal Henderson, and Paul Puryear (eds.) (1987). "Urban Politics and Urban Policy" in *The New Black Politics* (2nd ed., Part III). New York: Longman.

Villareal, Roberto E. and Norma Hernández (eds.) (1991). *Latinos and Political Coalitions*. New York: Praeger.

Wilson, William Julius (1978). *The Declining Significance of Race*. Chicago: University of Chicago Press.

PART V

Organization and Process

10

Citizenship and Economic Development

ELAINE B. SHARP
MICHAEL G. BATH

What is the role of citizens in the local economic development process? What forms of citizen involvement on developmental issues have been observed, and under which circumstances can activation of one or the other type be expected?

Because of its long-standing attention to political participation in general and its development of a wide variety of theories of participation, the political science discipline might be expected to contain a variety of ready responses to the above questions. There are, however, at least two obstacles to applying the voluminous political science literature on political participation to the issue of citizen involvement in economic development. First, most theories point to what it is that activates citizens rather than what predicts the direction (valence) of their activation or the quality and effects of their participation. Often, for example, citizen participation is assumed to mean mobilization for conflict; at the other extreme, some research only focuses narrowly on a single form of cooperative citizen action. Little if any attention is given to the development of theory to account not only for citizen activation but also for whether it is manifested in conflictual activity or collaborative activity with government. In this regard, much theory is likely to be of limited utility for a theory-praxis connection.

A second obstacle is Peterson's (1981) influential work, which suggests that citizen participation in the developmental sphere will tend to be limited, especially in comparison with citizen involvement in the distributional arena. According to Peterson (1981: 132), competition among cities for economic investment means that developmental matters are in the unitary interest of the city. Hence developmental policies tend to involve "highly centralized decision-making processes involving prestigious businessmen and professionals" with minimal conflict or citizen activation over the issue. If Peterson is correct, a theory of citizen involvement in local economic development is a contradiction in terms.

But there are counters to each of these problems. With respect to Peterson's thesis, there is substantial case study evidence of citizen mobilization around economic development issues, especially in the form of controversies over particular development policies and projects (Swanstrom, 1985; Jones and Bachelor, 1986; Fainstein et al., 1983). Given this evidence, Peterson's thesis becomes, rather than an obstacle to study of citizen involvement in economic development, an invitation to investigate factors that condition whether economic development is depoliticized or politicized, pluralistic or elitist.

The tendency of political science theories to focus on who participates or the conditions for political mobilization rather than the character and direction of citizen involvement is a more substantial problem. This chapter is directed toward a synthesis and adaptation of existing theories of citizen participation, designed to overcome this problem.

Theories of Political Participation

A menu of forms of citizen participation would include electoral involvement, protest and complaint activity, and various problem-solving behaviors. As Table 10.1 shows, each type of participation can also be categorized as having a more group-based and a more individualistic form.

A vast literature, incorporating a diversity of theories of participation, has accumulated on these phenomena. To determine which of the theoretical perspectives are most relevant to the local economic development context, it is important initially to determine which types of political participation are observable with respect to local economic development.

The case study evidence suggests that citizen participation with respect to economic development matters does run the gamut from electoral to protest or complaint to problem-solving activities. What is distinctive about it, however, is its group-based orientation. That is, local economic development often involves neighborhood organizations and other territorial associations protesting, lobbying, and even litigating against development projects that are viewed as threatening (Jones and Bachelor, 1986); citizens have also been involved in group-based electoral coalitions focused either on candidates with developmental policy agendas (Swanstrom, 1985) or on the use of direct democracy devices to tackle controversial developmental issues and policies (Caves, 1992). Collaborative problem solving or coproductive efforts in the economic development sphere have also been observed. These include various public-private partnerships or develop-

TABLE 10.1. Types of Citizen Participation

	Group Based	Individualistic
Electoral	Campaigning Direct democracy efforts (initiative, referendum)	Voting
Protest/complaint	Organized protest Neighborhood organization Lobbying/demand making Organizational litigation	Individual contacting and complaints
Problem solving	Community development corporations Group-based coproduction	Volunteer work

ment planning initiatives involving neighborhood organizations (Daykin, 1988).

It is not surprising that economic development inspires group-based rather than individualistic modes of participation. Unlike routine urban services, developmental projects tend to have broader impacts. This is because most urban service delivery decisions are quite "direct, daily, and locality-specific" (Yates, 1977), whereas economic development decisions have broader and longer-term consequences for the community as a whole, as well as spatially differentiated effects within the community.

A number of influential theories have been developed to account for the various forms of political participation that are relevant here. Broadly speaking, they can be divided into three types: (a) *psychosocial* theories, which emphasize individual attitudes and the social groupings that condition the development of individual attitudes, beliefs, and so on; (b) *rational calculus* theories, which assume that individuals are mobilized into group-based political action on the basis of their objective assessment of the impact of proposed policies or existing arrangements; and (c) *institutionalist* theories, which emphasize the importance of various institutional arrangements in either fostering or limiting citizens' access to governmental decision-making arenas. The following section briefly describes key theories of each type.

Psychosocial Theories

A variety of psychological predispositions, but most notably trust in government and a sense of personal efficacy, have been theorized as facilitators of political participation; similarly, a variety of social characteristics, such as income, age, social status, gender, and the like, have been hypothesized to have a functional relationship with one or another form of political participation. But perhaps the most substantial of the various theories of political participation are ones that draw upon both social characteristics and psychological predispositions.

For example, at least in its original formulations, *relative deprivation theory* suggested that political mobilization, in the form of destabilizing mass movements, is driven by an alienation phenomenon that is itself derived from a clash between expectations and social circumstances. Political revolution has been hypothesized to be most likely when an extended period of economic and social development is succeeded by a short period of sharp reversal, because expectations continue to rise even though perceptions of deprivation set in (Davies, 1962; Gurr, 1970). Although relative deprivation theory has generated a substantial body of empirical work on political upheavals across nation-states, it has not captivated the imaginations of those studying less revolutionary forms of political participation at the subnational level in the United States.

Perhaps the most important and basic theory of political participation is the *standard, socioeconomic model*. As articulated by Verba and Nie (1972), this theory suggests that the propensity for political participation is positively related to individual social status. Higher levels of participation are expected from higher income, better educated individuals not only because such individuals have more resources for political involvement but also because such individuals are more likely to have developed "such civic orientations as concern for politics, information, and feelings of efficacy, and these orientations in turn lead to participation" (Verba and Nie, 1972: 126). The standard, socioeconomic model has long been a staple in the study of voting for political candidates. It has also been shown to hold up with respect to voting on propositions (Magleby, 1984: 120).

The standard, socioeconomic model has been of much less utility in accounting for citizen-initiated contacting of city officials (Sharp, 1986). It has, however, been notably successful in predicting citizen involvement in various forms of cooperative, communal problem-solving activity. Verba and Nie (1972: 135), for example, found that the standard, SES model works better in predicting communal activity such as participation in group-based problem solving than for any other form of participation, and at least some of the research on neighborhood organizations in particular suggests that those of higher socioeconomic status participate more (Thomas, 1986; but see Haeberle, 1987). Similarly, research on coproduction and volunteerism supports the standard, SES model's contention that those at higher levels of social well-being are more likely to participate (Sharp, 1990: 93).

Rational Calculus Theories

Rational calculus theories are those that focus attention on conscious choices and self-interested action rather than upon the attitudinal structures and predisposing social contexts of psychosocial theories. For example, *expectancy theory*, which has been used to account for citizen volunteerism (Anderson and Moore, 1978), posits that individuals choose to participate or not based on their calculus of the desirability of various outcomes of participation, weighted by their probability of occurrence. Expectancy theory portrays the individual as a conscious calculator, choosing on the basis of a rational decision a course of action that will maximize utility. There have been only limited empirical applications of expectancy theory, however. In such applications, expectancy theory has generated relatively weak support (Miller, 1985).

With respect to economic development, rational calculus theory also supports the growing emphasis on property stakes, for these are an important basis for expected utility. A variety of theorists have begun to try to account for political mobilization through attention to such property stakes. Drawing largely upon the conceptual and functional difference between the "use value" and the "exchange value" that can be derived from property, various theorists have developed explanations of the genesis and direction of land use conflicts at the local level that are linked to differences in ownership and tenancy status (Logan and Molotch, 1987; Davis, 1991).

Perhaps the most important of the rational calculus theories, however, is Mancur Olson's (1971) *theory of collective action*. According to Olson's formulation, the mobilization of individuals into a group for pursuit of some collective purpose is problematic because of the free rider problem. To a potential group member, it is logical to assume that others in the group will bear the cost of providing the collective good, so there is no incentive for any individual to contribute. The provision of material benefits and other "selective incentives," available only to group contributors, however, can remedy the collective action problem by stimulating the free rider to act in a group-oriented way. The provision of these incentives to join entails certain costs, however, which some latent groups may not be able to cover. Thus the collective action problem and the costs involved in overcoming it act as constraints on the mobilization of groups.

Olson's thesis is consistent with, and helped to spawn, an entire genre of work on group mobilization in general and the activities of entrepreneurs who marshal the resources needed for selective incentives. Similarly, *resource mobilization* theory has been developed to account for the activation of citizens into mass movements and protest activities. In

contrast with earlier theorizing about social movements, which emphasized emotivist actors, impulsive dynamics, and the general psychology of
mass action, resource mobilization theory has moved "toward an examination
of rational motives, successful organizational strategies, and structural opportunities for mobilization" while emphasizing the entrepreneurial activity of
professional movement organizers (Capek and Gilderbloom, 1992: 33).

Hirschman's (1970) *participatory theory,* drawing upon a typology of
"exit, voice, and loyalty," is properly considered within the rubric of
rational calculus theory because it portrays participation as a form of
problem-solving activity and considers the strategic interrelationships among
the various choices. Many of the usual forms of political participation (contacting, protest, voting) are encompassed by Hirschman's "voice" category,
which incorporates attempts to solve problems by exerting pressure on
political or organizational leaders. "Exit," on the other hand, involves all
situations in which the individual responds to a problem by leaving (i.e.,
quitting an organization, leaving a community). "Loyalty" is the choice to
stay, even in the face of discontents, in the expectation that "someone will
act or something will happen to improve matters" (Hirschman, 1970: 78).

Hirschman's most important insight is the possibility that the existence
of one option can affect the choice of the others. For example, Hirschman
suggests the possibility that exit atrophies voice. That is, while exit is the
more dramatic response and might normally be expected to be the last resort
after voice has failed, the existence of an exit option may tempt some
individuals to leave before fully exhausting voice. And, the more readily
available is the exit option, the less sustained and widespread will be the
use of voice.

There have been some efforts to make empirical applications of Hirschman's
theory to political participation at the local level (e.g., Sharp, 1984). More
generally, however, the exit and voice options have been investigated
separately rather than being explicitly treated as trade-offs in the manner
suggested by Hirschman.

Institutionalist Theories

An important set of theories of participation place their explanatory
emphasis upon institutional arrangements that condition citizens' political
activation or quiescence. In the urban politics field, four sets of institutional arrangements have been the focus of particular attention: (a) metropolitan governmental organization, (b) reformed or unreformed institutions
of local government, (c) specific arrangements for citizen involvement in
local governance, and (d) public authorities and similar quasi-public

entities. Theory with respect to the first of these has, due largely to Tiebout's (1956) influential work, been an application of microeconomic theory to the metropolitan realm. Important extensions and applications of this theory have been attempted (Schneider, 1989), but the essential insights of the theory with respect to political participation are preserved, intact, in a *political economy* theory that serves as a challenger to metropolitan reform traditions. The political economist's theory predicts that political participation is greater in a fragmented metropolitan area than in a more centralized metropolitan area (Ostrom, 1984) because citizens will find the larger, more hierarchical organizations of a unified metropolitan government less accessible and less understandable.

Theory with respect to the impact of reform institutions on political participation is largely derived from interpretations of the logical implications of the reform movement. These include the presumption that citizen participation will, overall, be lower in reformed than in unreformed settings, because the depoliticization of local affairs through nonpartisanship and at-large elections short-circuits the avenues for mobilization of the public by politicians and because the reform institutions of merit selection and professional administration delegitimize citizen involvement on the administrative side of governance (Welch and Bledsoe, 1988).

In part because of the legacy of federally mandated citizen participation requirements and in part because of the need to counterbalance the unresponsive tendencies of reformism, cities have introduced a variety of organizational arrangements for citizen access to local government decision making— arrangements ranging from citizen boards to cablecasting of city council meetings to municipal ombudsmen (Sharp, 1990). In addition, increased use of the initiative and referendum at the subnational level has focused attention on these institutional arrangements of direct democracy (Caves, 1992). Theoretically, existence of institutional arrangements such as these should enhance citizen participation, particularly of the electoral and protest/complaint type. Moreover, some empirical research suggests that city governments' developmental policy activity is more responsive to fiscal stress in places where institutional arrangements for citizen access to development policymaking are more elaborate (Sharp, 1991).

Finally, the potential for citizen participation in economic development planning and decision making can be inhibited by the housing of development activities in quasi-public entities such as development boards or corporations or authorities. Various analysts have documented the ways in which such quasi-public entities can be used to insulate development activities from public awareness and involvement and to allow private interests to function virtually as a shadow government (Stoker, 1987; Hula, 1990).

Combinatorial Theory

Theories that draw insights from some combination of psychosocial, rational, and institutionalist perspectives and set forth explanations keyed to explicit interactions among some of these variables are very rare. One that draws from at least two of the theoretical perspectives is Jones's (1980) *need-awareness model*, which has become a very influential theory of citizen-initiated contacting of officials. Jones specifies that the propensity to contact is not simply a function of need for assistance, because objective need is politically inert without the knowledge, attitudes, and resources—that is, the political awareness—that psychosocial theorists have shown to be so important for political participation. On the other hand, while political awareness may in itself predispose upper-class citizens toward higher levels of system-affirming forms of participation such as voting, political awareness in itself is not a sufficient condition to mobilize citizens for instrumental forms of participation such as citizen-initiated contacting. Thus Jones concludes that a sufficient level of both need and political awareness are required to activate high levels of citizen contacting. Because need is inversely related to social well-being while political awareness is positively related to social well-being, the necessity for threshold levels of both predicts that citizen contacting would be highest in middle-class areas, a prediction consistent with Jones's observations in Detroit.

While Jones's need-awareness model has become a central theory for studying citizen contacting, it has not been applied to explanation of other forms of citizen participation. And, because citizen contacting is dominated by particularistic concerns and basic service delivery matters rather than the broader, longer-range issues of the developmental sphere, the relevance of Jones's theory has gone unaddressed with respect to the latter. Jones's theory will, however, be of some importance in this chapter as a basis for constructing a synthetic theory of developmental participation.

Limitations of Theories of Citizen Participation

The theories of participation outlined in the preceding section offer a variety of potentially powerful insights. These theories, however, are also problematic in several respects. First, the theories have not typically been synthesized, either in the sense of being tested against each other to demonstrate their relative merit in empirical application or in the sense of a deductive fusion of separate theories into a more comprehensive one. Second, the generalizability of the theories across a wide spectrum of forms of participation has not, with the exception of the standard socio-economic model, been broadly addressed.

Third, and perhaps most important for our purposes, while the various theories of participation may be useful in predicting the mobilization of citizens (and for diagnosing which types of citizens are more likely to be mobilized), they are not as immediately useful in accounting for the direction, or valence, that citizen participation takes. Many of the theories, such as relative deprivation theory, are based on the implicit assumption that the activation of citizens means mobilization for *conflict*—that is, citizen participation is treated as a struggle between citizens and government. Other theories, such as the standard, socioeconomic model, are apparently quite general, but, on closer inspection, the logic is more attuned to presumptions that participation is a *collaborative or system-affirming* act. Not surprising, such theories do better at explaining variation in behaviors such as voting and coproduction—both of which involve citizens in supportive, constructive, nonconflictual relationships with government.

What is really needed is a theory that can account not only for whether or not citizens are likely to be mobilized but also for the character of their relationship to government when mobilized. As Table 10.2 shows, consideration of these dual matters in the context of citizen involvement in local economic development generates four possible outcome types. If citizens are not activated but are positive in their orientation toward government's economic development agenda and activities, they are either satisfied or, like Hirschman's (1970) category of "loyalists," apathetically waiting for someone else to take action to deal with problems. If citizens are not activated even though they find government's economic development activities and agenda objectionable, they may be characterized as alienated. They may also be candidates for exit, the ultimate withdrawal from the community. Citizens who are mobilized to participate on the basis of a positive view of local economic development policies are likely collaborators, available for constructive involvement in planning and coproduction. On the other hand, citizens who are mobilized to participate but who have a negative view of government's activities with respect to economic development will presumably be engaged in any of a number of conflictual activities, such as protest, campaigning, or litigation to block development projects that have been fostered by local government.

Viewed in this way, the limitations of most theories of citizen participation are stark. They tell us the conditions under which citizen mobilization can be expected but not necessarily the character of that citizen activity. A theory that might be helpful in accounting for variation across all four cells of Table 10.2 would presumably draw upon the best of those theories of citizen participation; but it would need also to be based upon a logic that considers the activation of positive versus negative evaluations of government's economic development initiatives.

TABLE 10.2. Citizen Roles in Local Economic Development

Citizen Orientation Toward Government's Economic Development Agenda	Citizen Mobilization	
	Yes	No
Positive	Collaboration, coproduction	Loyalty
Negative	Conflict	Alienation, exit

A theory that does this can be constructed on the basis of an expanded version of Jones's (1980) need-awareness theory. As noted above, this theory explicitly draws upon elements from two of the categories of theory defined in this chapter—that is, the psychosocial and the rational calculus. By specifying that citizen activation is an interactive effect of need and political awareness, Jones's theory incorporates most of the rich insights that can be found in rational calculus theories (which emphasize objective stakes and strategic calculations, or the needs of citizens) and psychosocial theories (which emphasize attitudinal predispositions and social resources, or the awareness of citizens).

Two additions to Jones's theory are necessary. One is acknowledgment of the set of institutional arrangements and resources that have a bearing on the ability of entrepreneurs to mobilize citizens around developmental issues. These institutional arrangements can be viewed as mediating the effects of need and awareness on political mobilization. For example, even politically sophisticated citizens will be more difficult to mobilize when developmental decision making is hidden away in inaccessible venues such as development authorities; and need for development activity will be more easily translated into political activation in communities that have a broad range of institutionalized devices for citizen access, including the potentially powerful tools of direct democracy.

Second, need-awareness theory must also be adapted so that it can account for the direction, or valence, of citizen activation rather than simple mobilization. This can be done with elaboration on the need component of the model. As applied to citizen mobilization in the developmental arena, need might be understood to refer to the stakes that particular types of citizens have in government's developmental policy initiatives. Peterson (1981) characterizes these stakes as collective in nature and tied to the economic competitiveness of the city. That is, economic development activity is alleged to be in the unitary interest of the city. When the community is economically distressed, everyone has a strong stake in making the community more economically competitive;

when the community is already prosperous, everyone's stakes in and commitment to further development initiatives are more limited.

Economic development initiatives also have a number of individually differentiated impacts, however. That is, there are some negative consequences of growth in economic activity—such as congestion, pollution, and displacement—and these negative consequences, which are typically but not exclusively in the form of threats to use value, fall unevenly across the urban landscape and across different individuals and groups (Feagin, 1988).

Thus, in the development sphere, unlike the allocation sphere for which need-awareness theory was first developed, there are two dimensions of need—collective or community need (for jobs, growth, and prosperity) and individual need (for insulation from the negative consequences of growth). Recognition of the potential conflicts between one's need as a member of the collectivity and one's individual need provides grounds for predicting the specific forms that citizen participation in economic development may be expected to take.

The extended version of need-awareness theory, which we might call *developmental participation theory*, offers such predictions (see Figure 10.1). Developmental participation theory posits that, (a) if political awareness is sufficiently high, and (b) contingent upon the mediating effects of institutional arrangements such as reformism, governmental fragmentation, citizen participation arrangements, and quasi-public development bodies, (c) the propensity for and character of citizen participation in economic development is a function of community and individual needs in the developmental sphere. The theory also suggests the possibilities for movement from one outcome to another, depending in large part upon the way in which local leaders attempt to redefine citizens' perceptions of the city's developmental needs and the costs and benefits to individuals of particular development projects.

Specifically, the theory stipulates that conflictual modes of participation (i.e., protest or other efforts to block plans and projects) are, other things being equal, likely wherever and for whomever use values are negatively affected by development initiatives. In economically distressed communities, city leaders faced with such conflict face two possibilities. They can attempt to demobilize those citizens involved in the conflict, typically through the use of nonresponsiveness tactics that exhaust the challenging groups' resources. This leads to alienation or exit on the part of those defeated and demobilized. Alternatively, city officials can respond in ways designed to defuse perceptions of highly negative individual consequences of development initiatives, either by negotiating actual changes in the initiatives or through symbolic manipulation of perceptions. If successful, this transformation can lead to collaboration based upon a congruence of

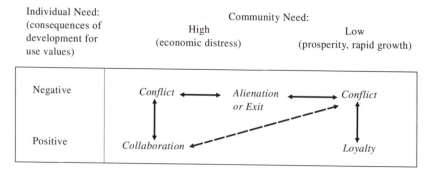

Figure 10.1. Developmental Participation Theory: Community Need, Individual Need, and Participatory Outcomes

individual and community needs. As we will see below, the history of citizen participation in economic development in Pittsburgh is illustrative of these sequences.

In relatively prosperous cities, economic development may not be viewed as an imperative at any cost. If there are individuals whose use-value stakes stand to be harmed by the prevailing direction of development policy and projects, conflictual modes of participation can emerge. The theory outlined in Figure 10.1 suggests that the prospects for movement away from such conflict over development are different in prosperous settings than in economically distressed communities.

Officials facing development-focused conflict in prosperous communities are facing three possibilities. Like their counterparts in economically distressed communities, they can use the resources at their disposal to fend off and insulate themselves from challengers. As in economically distressed communities, this strategy, if successful, leads to either alienation or the "exit" of disgruntled residents. Also, like their counterparts in economically distressed communities, they can respond in ways designed to alter perceptions that development initiatives create negative consequences for some individuals, either by negotiating actual changes in the initiatives or through symbolic appeals. Unlike the situation in economically distressed communities, this approach, if successful, is likely to transform a mobilized citizenry into a passive but optimistic citizenry (i.e., the "loyalty" outcome) rather than to transform conflict into collaboration. This is because defusing concerns about impacts on individual use values removes the only form of perceived need with respect to economic development. Absent any beliefs about the urgency of economic development for the community as a whole and satisfied about the protection of their individual use-value stakes, citizens have no reason for active participation with respect to development.

The third alternative can, if successful, transform conflictual modes of citizen participation into collaboration on economic development. Unlike their counterparts in economically distressed communities, however, officials in prosperous communities must manage *two* processes to accomplish this transformation. They must not only neutralize perceptions of negative use-value impacts from development projects but also convince citizens that economic development is a critical imperative despite the community's relative prosperity.

The various transformations hypothesized in the foregoing section can be illustrated with case examples drawn from the burgeoning literature on the politics of economic development. The long history of economic development efforts in Pittsburgh, Pennsylvania, provides illuminating detail on transformation from conflict to collaboration in an economically distressed community. By contrast, examples from Gainesville, Florida, and Lawrence, Kansas, illustrate some of the dynamics of conflictual participation over development in relatively prosperous communities.

Economic Development in Pittsburgh: Conflict to Collaboration

Pittsburgh's period of strident conflict over development involved the mobilization of neighborhoods in the 1960s against urban renewal projects that threatened to displace lower-income residents. In response to this, the most extreme version of negative use-value impacts, conflictual forms of citizen participation, emerged:

> In response, however, neighborhoods organized to stop urban renewal projects or to change their focus from clearance to rehabilitation. Neighborhood groups learned the routines of organizing, protest, confrontation, and lawsuits. (Sbragia, 1989: 108)

Transformation to a collaborative model began during the Flaherty administration but was solidified in the 1980s during the mayorship of Richard Caliguiri, who reoriented the city's developmental policy toward positive contributions to the neighborhoods. During the Caliguiri administration, neighborhood-based associations were invited into the developmental partnership that had long existed between local government and the business community; and substantial developmental resources, in the form of Community Development Block Grant funds and Urban Development

Action Grant monies, were targeted to neighborhood revitalization. In contrast with earlier periods of emphasis on the needs of the downtown, three-fourths of the federal development funds in Pittsburgh were targeted to the neighborhoods in the early 1980s (Sbragia, 1989: 109). City-government has also, with the support of foundations,

> encouraged and helped create community development corporations. Five such CDCs have become so visible and active that they are dominating the neighborhood agenda; economic development rather than provision of social services or advocacy is now viewed as the neighborhood issue deserving of highest priority. (Sbragia, 1989: 110)

The result of all of this is a major transformation from the earlier, conflictual mode. Sbragia (1990: 59) describes the "strikingly consensual nature of redevelopment in Pittsburgh in the 1980s" and explains that a key to this transformation is change in developmental policy strategy: "Rather than transforming land use, it hopes to restructure both the economic base and the social structure of the city" (Sbragia, 1990: 60). Such a move away from a land redevelopment focus is a prime example of the kind of change that can defuse the conflicts rooted in negative use-value impacts.

On the other hand, a focus on restructuring the economic and social base of the community also has the potential for highlighting negative impacts on certain segments of the community, once again yielding conflict. In Pittsburgh, where restructuring from a manufacturing to a postindustrial, corporate center means major dislocations for displaced workers, the potential for conflictual mobilization is very real.

Instead, unions and labor activists, in alliance with community organizations, have mobilized to participate in ways that push for their interests within a collaboration with government officials. This has largely occurred through the Tri-State Conference on Steel, which initially focused on plant closing situations such as Nabisco's proposal to close its Pittsburgh plant in 1982. In response to this, Tri-State orchestrated a pressure campaign against Nabisco, including an eminent-domain takeover proposal, and enlisted the support of Mayor Caliguiri in the campaign. As a result, Nabisco "capitulated to public pressure and kept the plant open, maintaining all jobs" (Fitzgerald and Simmons, 1991: 518). Tri-State was also successful in convincing Pittsburgh and seven other municipalities to join with labor representatives, community activists, and small businesses in a regional planning body, the Steel Valley Authority (SVA), whose mission is to head off manufacturing disinvestment. In contrast with SVA's agenda, city leadership in the Pittsburgh area is generally more oriented toward downtown businesses and major institutions representative of the postindustrial sector, such as the University of Pittsburgh and Carnegie Mellon

University. Nevertheless, the collaboration between city officials and community and labor activists in the SVA has led to at least some concessions in the city's development strategy (Fitzgerald and Simmons, 1991: 520). These concessions are likely to be crucial in neutralizing arguments that Pittsburgh's development strategy is leaving particular segments out in the cold, and hence crucial to maintaining the unusually collaborative form that development participation has taken in Pittsburgh.

Economic Development in Prosperous Communities

Pittsburgh is an aging, "Rust Belt" city in which the community's need for economic development initiatives is clear and compelling. By contrast, communities such as Gainesville, Florida, and Lawrence, Kansas, are blessed (and sometimes cursed) with substantial growth and, although far from uniformly wealthy, are relatively prosperous. In such communities, economic development activities that threaten residents' definition of the community's quality of life can lead to highly conflictual modes of citizen participation. As this section suggests, such conflict can be relatively durable in such settings, and transformation to enduring, collaborative modes of participation is more difficult.

In Gainesville, Florida, for example, a coalition of antigrowth activists emerged to challenge policies that were viewed as contributing to urban sprawl, congestion, pollution, and a general degradation of the quality of life in this university town. In a series of strident elections in the 1980s, pro- and antigrowth factions contended against each other, and the university and the chamber of commerce were implicated in a scandal over alleged efforts by a chamber leader to discredit the antigrowth activities of two faculty members. Growth management strategies, which link new development to the placement of infrastructure and protect environmentally sensitive areas, have apparently defused environmentalists in the antigrowth coalition, but neighborhood activists in the coalition are not satisfied with such compromise strategies because their use values are threatened by increased density even if there are environmental preservation zones elsewhere (Vogel and Swanson, 1989).

In Gainesville, therefore, citizen participation with respect to economic development is likely to remain largely conflictual. The extent of the stalemate is reflected in efforts to change the participation dynamics by manipulating the institutional arrangements for development decision making:

> To the antigrowth activists, it appeared that the growth machine had decided that if it could not run the city, it would dismantle it. The state legislative delegation passed local bills that stripped selected functions from the city. A separate library

district was created. Efforts continued to set up an airport authority, a joint
city-county planning agency, consolidated police protection under the county
sheriff, and an independent utility authority. (Vogel and Swanson, 1989: 74)

Like Gainesville, Lawrence, Kansas, is a university community that has
been experiencing substantial growth and episodes of conflictual participation stemming from residents' objections to the negative impacts of
development projects on their neighborhoods. One of these conflicts
focused on "the Bluffs, a rocky hill of five acres described by Pre-
servationists as one of Lawrence's 'most beautiful natural resources' "
(Schumaker, 1991: 92). In response to a local developer's proposal to build
apartments and offices on the site, city commissioners rezoned the land from
its original, single-family designation. As a result, the adjacent neighborhood
organized and filed suit against the city; later, the developer also filed suit
when the city balked on a building permit on technical grounds.

In this case, conflict was transformed into quiescence when lawyers
representing all sides of the issue came up with a settlement that defused
the negative impacts on the neighborhood:

> The developer received his building permit and an additional change in
> zoning, permitting more extensive office developments. The neighbors were
> "buffered" from the densest developments on the site and were sheltered from
> increased traffic by an agreement to build a new road providing direct access
> to new developments (Schumaker, 1991: 93).

As in many other high-growth communities, however, settlement of one
controversy buys only a limited period of quiescence before the emergence
of yet another. In Lawrence, residents of one or another neighborhood have
mobilized over a variety of development projects, over proposals to "downzone"
particular neighborhoods to preserve their single-family character, and many
other projects. Each of these controversies is ultimately settled, sometimes
to the satisfaction of residents challenging development projects, sometimes not, and sometimes in ways that simply defer the conflict. But none
of the controversies appears to serve as a vehicle for systematic transformation from repeated use-value conflicts to a sustained, collaboration
between citizens, private developers, and city government based on a
shared vision of the community's developmental needs.

Conclusion

Although there are many different theoretical perspectives on citizen
participation, relatively few are both comprehensive and specifically tai-

lored toward understanding citizen involvement in local economic development. Theory that is designed to account for the style or valence of citizen participation as well as for political activation itself is even more notable by its absence. Ironically, Jones's (1980) need-awareness theory, which was *not* designed with developmental participation in mind, offers perhaps the most useful springboard for a synthetic theory, incorporating the insights of psychosocial and rational calculus theories.

As this chapter shows, such a synthesis highlights the importance of two aspects of need with respect to economic development—the community's need for development and the individuals' need to be protected from negative consequences of development initiatives. It also suggests that the dynamics of citizen participation in economic development are substantially different in economically distressed and economically advantaged communities. Conflict over economic development is likely to be more sustained and more difficult to transform into collaboration in relatively prosperous communities, not simply because of the prevalence of higher social status citizens with more "political awareness" resources but because of the interaction between such "awareness" resources and high levels of perceived need for protection from negative use-value impacts, untempered by perceptions that development is a compelling need for the community.

References

Anderson, John C. and Larry F. Moore (1978). "The Motivation to Volunteer." *Journal of Voluntary Action Research* 7 (Summer-Fall): 120-129.

Capek, Stella M. and John I. Gilderbloom (1992). *Community Versus Commodity*. Albany: State University of New York Press.

Caves, Roger W. (1992). *Land Use Planning: The Ballot Box Revolution*. Newbury Park, CA: Sage.

Davies, James C. (1962). "Toward a Theory of Revolution." *American Sociological Review* 27 (February): 5-19.

Davis, John Emmeus (1991). *Contested Ground*. Ithaca, NY: Cornell University Press.

Daykin, David S. (1988). "The Limits to Neighborhood Power: Progressive Politics and Local Control in Santa Monica" in Scott Cummings (ed.), *Business Elites and Urban Development*. Albany: State University of New York Press.

Fainstein, Susan, Norman Fainstein, Richard Child Hill, Dennis Judd, and Michael Peter Smith (1983). *Restructuring the City*. New York: Longman.

Feagin, Joe (1988). "Tallying the Social Costs of Urban Growth Under Capitalism: The Case of Houston" in Scott Cummings (ed.), *Business Elites and Urban Development* (pp. 205-234). Albany: State University of New York Press.

Fitzgerald, Joan and Louise Simmons (1991). "From Consumption to Production: Labor Participation in Grass-Roots Movements in Pittsburgh and Hartford." *Urban Affairs Quarterly* 26 (June): 512-531.

Gurr, Ted Robert (1970). *Why Men Rebel*. Princeton, NJ: Princeton University Press.

Haeberle, Steven (1987). "Neighborhood Identity and Citizen Participation." *Administration & Society* 19 (August): 178-196.

Hirschman, Albert O. (1970). *Exit, Voice, and Loyalty.* Cambridge, MA: Harvard University Press.

Hula, Richard C. (1990). "The Two Baltimores" in Dennis Judd and Michael Parkinson (eds.), *Leadership and Urban Regeneration* (pp. 191-215). Newbury Park, CA: Sage.

Jones, Bryan (1980). *Service Delivery in the City.* New York: Longman.

Jones, Bryan and Lynn Bachelor (1986). *The Sustaining Hand.* Lawrence: University Press of Kansas.

Logan, John and Harvey Molotch (1987). *Urban Fortunes: The Political Economy of Place.* Berkeley: University of California Press.

Magleby, David (1984). *Direct Legislation.* Baltimore: Johns Hopkins University Press.

Miller, Lynn E. (1985). "Understanding the Motivation of Volunteers: An Examination of Personality Differences and Characteristics of Volunteers' Paid Employment." *Journal of Voluntary Action Research* 14 (April-September): 112-122.

Ostrom, Elinor (1984). "Metropolitan Reform: Propositions Derived from Two Traditions" in Charles Levine (ed.), *Readings in Urban Politics: Past, Present and Future* (2nd ed., pp. 329-347). New York: Longman.

Olson, Mancur (1971). *The Logic of Collective Action.* Cambridge: Harvard University Press.

Peterson, Paul (1981). *City Limits.* Chicago: University of Chicago Press.

Sbragia, Alberta (1989). "The Pittsburgh Model of Economic Development: Partnership, Responsiveness, and Indifference" in Gregory Squires (ed.), *Unequal Partnerships* (pp. 103-120). New Brunswick, NJ: Rutgers University Press.

Sbragia, Alberta (1990). "Pittsburgh's 'Third Way': The Nonprofit Sector as a Key to Urban Regeneration" in Dennis Judd and Michael Parkinson (eds.), *Leadership and Urban Regeneration* (pp. 51-68). Newbury Park, CA: Sage.

Schneider, Mark (1989). *The Competitive City: The Political Economy of Suburbia.* Pittsburgh: University of Pittsburgh Press.

Schumaker, Paul (1991). *Critical Pluralism, Democratic Performance, and Community Power.* Lawrence: University Press of Kansas.

Sharp, Elaine B. (1984). "'Exit, Voice, and Loyalty' in the Context of Local Government Problems." *Western Political Quarterly* 37 (March): 67-83.

Sharp, Elaine B. (1986). *Citizen Demand-Making in the Urban Context.* University, AL: University of Alabama Press.

Sharp, Elaine B. (1990). *Urban Politics and Administration.* New York: Longman.

Stoker, Robert P. (1987). "Baltimore: The Self-Evaluating City?" in Clarence Stone and Heywood Sanders (eds.), *The Politics of Urban Development* (pp. 244-268). Lawrence: University Press of Kansas.

Swanstrom, Todd (1985). *The Crisis of Growth Politics.* Philadelphia: Temple University Press.

Thomas, John C. (1986). *Between Citizen and City.* Lawrence: University Press of Kansas.

Tiebout, Charles (1956). "A Pure Theory of Local Expenditures." *Journal of Political Economy* 64: 416-424.

Verba, Sidney and Norman Nie (1972). *Participation in America.* New York: Harper & Row.

Vogel, Ronald and Bert Swanson (1989). "The Growth Machine Versus the Anti-Growth Coalition: The Battle for Our Communities." *Urban Affairs Quarterly* 25 (September): 63-85.

Welch, Susan and Timothy Bledsoe (1988). *Urban Reform and Its Consequences.* Chicago: University of Chicago Press.

Yates, Douglas (1977). *The Ungovernable City.* Cambridge: MIT Press.

Suggested Readings

Coulter, Philip B. (1988). *Political Voice: Citizen Demand for Urban Public Services.* Tuscaloosa: University of Alabama Press.

DeLeon, Richard E. and Sandra S. Powell (1989). "Growth Control and Electoral Politics: The Triumph of Urban Populism in San Francisco." *Western Political Quarterly* 42: 307-331.

Frieden, Bernard J. and Lynne B. Sagalyn (1989). *Downtown, Inc.: How America Rebuilds Cities.* Cambridge: MIT Press.

Gottdiener, Mark (1987). *The Decline of Urban Politics.* Newbury Park, CA: Sage.

Haeberle, Steven H. (1988). "Community Projects and Citizen Participation: Neighborhood Leaders Evaluate Their Accomplishments." *Social Science Quarterly* 69: 1014-1021.

Huckfeldt, Robert (1986). *Politics in Context: Assimilation and Conflict in Urban Neighborhoods.* New York: Agathon.

Hutcheson, John and James Prather (1988). "Community Mobilization and Participation in the Zoning Process." *Urban Affairs Quarterly* 23: 346-368.

Jezierski, Louise (1990). "Neighborhoods and Public-Private Partnerships in Pittsburgh." *Urban Affairs Quarterly* 26: 217-249.

Logan, John and Harvey Molotch (1987). *Urban Fortunes: The Political Economy of Place.* Berkeley: University of California Press.

Logan, John and Gordana Rabrenovic (1990). "Neighborhood Associations: Their Issues, Their Allies, and Their Opponents." *Urban Affairs Quarterly* 26: 68-94.

Nyden, Philip W. and Wim Wiewel (1991). *Challenging Uneven Development: An Urban Agenda for the 1990s.* New Brunswick, NJ: Rutgers University Press.

Saltman, Juliet (1990). *A Fragile Movement: The Struggle for Neighborhood Stabilization.* Westport, CT: Greenwood.

Stone, Clarence (1989). *Regime Politics: Governing Atlanta, 1946-1988.* Lawrence: University Press of Kansas.

Stone, Clarence and Heywood Sanders (1987). *The Politics of Urban Development.* Lawrence: University Press of Kansas.

11

Technology Transfer and Economic Development

JULIA MELKERS
DANIEL BUGLER
BARRY BOZEMAN

"Competitiveness" has been the battle cry of the 1980s. Since late in the Carter administration, the federal and state governments have attempted a variety of legislative and regulatory actions designed to stimulate domestic economic growth. At the federal level, these actions have primarily sought to increase the transfer of scientific and technical knowledge from national labs to private industry (Weingarten, 1989). State governments have mirrored federal efforts, implementing programs that attempt to stimulate technology-based economic development (Eisinger, 1988). While the number of laws and programs have multiplied, the process of moving science to market has proven to be difficult at best.

In 1988 the United States spent $64 billion creating new scientific and technical information, with national labs performing much of the nation's basic research (Ballard et al., 1989). Although federal resources *generate* many new scientific and technical discoveries, federal agencies are not the *users* of this knowledge. The gap between producers of new knowledge and the ultimate users, private industry, has generated a spate of laws and programs to speed the transfer of technology from the labs to the market (Lambright, 1979). As states jump deeper into economic development activities, they are increasingly instituting programs to attract, nurture, and develop high-technology industries (Eisinger, 1988; Schmandt and Wilson, 1990; Fosler, 1988). By the late 1980s, 43 states had at least one program encouraging technological innovation (Minnesota Department of Trade and Economic Development, 1988). In all, more than 250 technology-based economic development programs were instituted, spending over half a billion dollars (Osborne, 1990).

At the root of these programs is an implicit link between technology transfer and economic development. *Technology transfer* may be broadly defined as the transfer of a technology, technique, or knowledge that has been developed in one organization to another, where it is adopted and used. Economic development is generally understood to be an increase in the standard of living as represented by increases in employment and per capita income. The adoption of a technology by a firm or a number of firms does not necessarily lead to the creation of new jobs and economic growth in a region. So why are so many government agencies attempting to use technology transfer as a tool to spur economic development?

The answer lies in technical change literature and innovation theories. It is difficult to show a direct causal link between technology transfer and economic development, but technical change literature and innovation studies emphasize the role of new technology in economic growth. Governments have used the idea of this link to establish programs designed to encourage innovation with the assumption that technology and subsequently economic growth will result.

This emphasis on innovation to spur economic development is not new. A widely accepted view is that new technology is an early link in a chain of activities that lead to a reworked industrial structure (Hicks, 1988). This view holds that technology is catalytic and can be expected to stimulate the rise of entirely new industries. From the beginning of the 1900s, innovation has been seen as an "engine of economic growth" (Schumpeter, 1950).

In Schumpeter's work, innovation cycles drive the economy. The beginning of the cycle is marked by new inventions that are adopted by entrepreneurial firms; as more firms adopt the technology, economic growth occurs. Eventually the firms mature and the technology becomes increasingly outdated and the economy declines. A new cycle begins with the development of new technologies.

Innovation theories, however, do not directly address economic development. Rather, they are concerned with the macro level efforts of technology, technical change, and innovation on the economy (Cohen and Zysman, 1987; Dosi et al., 1988; Landau, 1989; Mowery and Rosenberg, 1989; Tushman and Moore, 1982). What government policy has done is to use these theories in their most basic sense to establish a justifiable argument for the establishment of technologically based programs. Understanding of this material and this link will aid in understanding of the assumptions behind technology-based economic development.

Government policies designed to address technology transfer will succeed or fail at the organizational level. Successful technology transfer requires transfer of knowledge across disciplines, professions, industry sectors, regions, and/or societies (Reisman, 1989). Thus it is an organizational and cultural process as well as a knowledge transfer process. We present a contingency model of technology transfer and a framework for assessing the effectiveness of technology transfer efforts.

This chapter brings together theories of the innovative process and the role of technology in an attempt to link technology-based economic development initiatives to a theoretical base. We begin by describing the process of innovation and the various assumptions about the innovative process. Both technology transfer and the innovative process are not well understood. Therefore we spend some time familiarizing the reader with technology transfer and its conceptual base. We then take a more practical turn and review technology transfer initiatives at the federal level and provide examples of state-level efforts. We conclude with some thoughts on the relationship between technology transfer and economic development.

The Innovative Process:
Theories of Innovation

A key assumption from the competitiveness debate is that science and technology fuel economic growth. It is derived from work by Schumpter, Solow, and Denison, which found causal relationships between innovation and economic development.[1] Solow was one of the first to find that these two inputs could not account for growth in the U.S. economy.[2] But the residual in the model tested correlated highly with U.S. growth (Stoneman, 1987). The residual included a number of factors that affect U.S. growth, but Solow labeled them "technological change" (Papadakis, 1990). In a study using a more refined model, Denison found the residual—advances in knowledge—accounted for 57% of the growth in productivity.

Rate of return studies have provided lawmakers with a powerful rationale for the support of efforts to increase R&D spending and to enhance technology transfer activities. They support explanations of the innovation process that are relatively simple and linear (Burton, 1989). (See Figure 11.1.) If science is the bedrock of economic growth, then the transfer of findings to industry should guarantee long-term growth.

Recent work on the innovation process suggests that the process is anything but linear. The relationship of science to technology and technology to economic growth is complex, interactive, and iterative (Tornatzky and Fleischer, 1990). Many factors affect the likelihood that a lab will develop scientific breakthroughs. Another set of factors affect the likelihood that a new technology will be successfully developed. A successfully developed new technology does not necessarily succeed in the market-

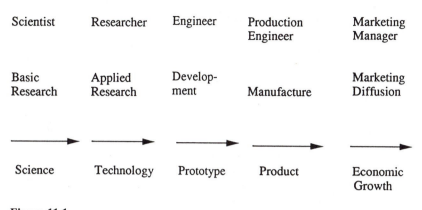

Scientist	Researcher	Engineer	Production Engineer	Marketing Manager
Basic Research	Applied Research	Develop- ment	Manufacture	Marketing Diffusion
Science	Technology	Prototype	Product	Economic Growth

Figure 11.1.
SOURCE: From Burton (1989).

place (Mansfield, 1982). All it offers are technological opportunities, which must then be matched to market demand in the economic system (Freeman, 1983). Thus the economic system provides the incentives for new innovations and influences the rate of innovations and the characteristics that new innovations will have. Ultimately, demand in the economic system determines the likelihood that a technological opportunity will reach commercial success (Mansfield, 1982).

The transformation of an innovative success to a market success is determined at the level of the firm. Information about new technological opportunities, and about new market demands, is filtered through the firm (Willinger and Zuscovitch, 1988). Constraints on the available technology and constraints on the firm will determine technological trajectories available to the firm (Dosi et al., 1988); the nature of demand and the markets, along with factors in the firm, will then determine which markets the firm seeks to exploit with a new innovation. The entire process is characterized by uncertainty, trial, and error (Kline and Rosenberg, 1986).

Adopting an innovation does not mean the firm can capitalize on it. Its ability to do so depends on its technical capacity, organizational ability, and knowledge of the market. What does this mean for policies designed to nurture technology transfer to industry? First, that simply transferring technology does not necessarily imply economic benefits. Policies need to address the uncertainty prevalent in decisions to adopt an innovation. Firms need technical capacity, suitable organizational structures and processes amenable to innovative behavior, and a way to analyze and understand market signals. In the next section of the chapter, a conceptual framework of technology transfer is developed.

Conceptualizing Technology Transfer

Like the innovation process, technology transfer is not well understood:

> It is uncertain exactly what constitutes a transfer. Does it occur with the awareness by one organization of the knowledge products available in other organizations? Or is there a commercial requirement for the term to have meaning? . . . [V]arious models of innovation . . . offer descriptions in process terms as if technology transfer is a single phenomenon . . . [T]his is not a description of one process but is the interface between at least two organizations that often are very different . . . [T]hus, dissemination and adoption of knowledge products operate at the boundary of the organization as the organization adapts to the environment. (Kingsley, 1990, p. 6)

Simply put, technology transfer is a complicated process involving several different variables and attributes. Instances of technology transfer vary according to differences in characteristics of the transfer agent, transfer type, transfer product attributes, the transfer mechanism, attributes of the transfer recipient, market contingencies (Bozeman and Fellows, 1988), and the boundaries between deployers and users (Tornatzky and Fleischer, 1990).

Different laboratories and different transfer agents require different approaches to technology transfer (Bozeman and Fellows, 1988). There is not only variance between the way that public and private laboratories approach technology transfer, but there is also a great deal of variance among government labs. Similarly, there is a difference in the preestablished or even nontechnological relationships between transfer organizations (Tornatzky and Fleischer, 1990). Some organizations have historically stable relationships that result in an interconnectedness between the transfer players.

The type of technology transfer, whether it is a transfer of scientific knowledge, physical technology, technological design, or process, affects the nature of the transfer. The attributes of the transfer product itself are important. The closer that a product is to commercialization as well as its potential breadth of application have a great deal to do with not only the nature of the transfer but also its likelihood of success.

Other, less predictable and uncontrollable attributes of the technology transfer process contribute both to the difficulty in its conceptualization as well as to the assessment of its effectiveness. The technology developers or transfer agents have no mechanism for evaluating the likelihood of any particular transfer recipient's ability and likelihood to develop a commercially viable product from a quality technology transferred by a lab. Finally, a transfer may result in what seems to be a quality product. The market for that product, as well as its life cycle and whether or not there are more affordable alternatives, however, have a great deal to do with whether the technology transfer will result in commercialization (Bozeman and Fellows, 1988).

Measuring the effectiveness of technology transfer is as difficult as conceptualizing the process itself. Successful technology transfers depend on a number of factors. As described above, the organizations involved usually have different goals and each case of technology transfer has unique qualities. This confounds the ability to assess the effectiveness of these efforts. The product transferred can have different attributes; the means of transfer differ. And there are different market contingencies that the transfer product addresses. The transfer process itself requires a significant amount of an organization's resources (time, money, opportunity costs). Because of this, successful transfers require a high level of commitment by both organizations.

Another important component in assessing technology transfer success is its time line. Scientific and technological development is a lengthy process. Often, its time line is unknown or uncertain. The length of time required for one product to reach commercialization may not be the same as another, even similar, product. The factors that make each transfer process unique make it that much more difficult to assess. Assessment is important, however, especially in times where accountability and justification of public support for science are highlighted. Most policy processes operate within shorter time horizons. These are usually attributed to electoral cycles and the short-term incentives and demands within the political arena (Barke, 1990). The focus of many state-level high-technology efforts is on quick results. This is not consistent with the long-term nature of the innovation process and the uncertain time line of technology transfer.

Most technology transfer efforts, especially those at the state level, are fairly new. There are "numerous technical difficulties associated with the newness of most programs, the absence of standardized effectiveness measures, the lack of data collection on program impact and the difficulty of attributing causality in such data where it exists" (Watkins and Wills, 1986). The need to justify federal and state support for technology-based economic development programs ensures that improved evaluation of these types of programs will exist in the future. The next section describes some of these technology transfer initiatives at both the federal and the state levels.

Theory in Practice:
Technology Transfer Legislation

The last ten years have seen significant changes in the way that government perceives and deals with research and development activities. The

1980s were a pivotal time period where there was a distinctive shift toward cooperative efforts in research and development. Various U.S. federal government and state policy initiatives have reflected the technical change and innovation theories described earlier in the form of technology transfer legislation. These policies focus not only on enhancing the linkages between government and industry but also on decreasing the impediments to those linkages. The technical change approach described in a previous section inspired three main governmental responses to the perceived economic crisis or lack of competitiveness. The initial response is to increase research and development and science and technology activities; second is an increase and improvement of the transfer of knowledge; third is the effort to remove impediments to innovation and the innovative process.

The problem is that, while the technical change and innovation theories suggest a link between technology, technology transfer, and economic growth, it is not a well-defined link. The technology transfer policies are in fact using a basic input-output innovation model based on a linear model of innovation. Innovation, however, is not a linear or straightforward process. What these policies fail to address are the complexities and uncertainties associated with technological change and economic growth.

This section reviews some of the major technology transfer efforts, including technology transfer legislation, followed by a discussion of more micro-level (state-level) responses. Implicit in all of these technology-based economic development policies or programs is some adherence to technical change theory and the linear model of innovation. The technical change literature embodies the innovation process and operates under the assumption that technical change brings about economic progress. As discussed above, the traditional (linear) innovation model oversimplifies the transfer of science to technology, technology to products, and products to market success and economic growth. Thus lawmakers are led to believe that the mere addition of more science or more technology will lead to faster growth rates, when in fact the problems in a particular area may lie elsewhere.

One way of conceptualizing technology transfer efforts is in terms of their underlying purposes—appropriability, dissemination, and knowledge use (Ballard et al., 1989). The appropriability model of technology transfer was most apparent from 1945 to the late 1950s. It emphasized the importance of producing high-quality R&D to ensure its use and derives in many ways from Vannevar Bush's report *Science—The Endless Frontier* (1954) that dealt with the role of postwar science in the United States.

The dissemination model focuses on the need to disseminate information and technology to potential users (Ballard et al., 1989). It is based on the realization that, if better dissemination mechanisms are available to link users with producers, improved opportunities for technology transfer would emerge. Programs of the 1950s-1970s embodied this model such as the Department of Commerce's Office of State Technical Services Program and the Intergovernmental Science, Engineering, and Technology Advisory Panel (Doctors, 1981).

Finally, the knowledge utilization model expands on the dissemination model by changing the focus of technology transfer to the relationships between its various components. It suggests an active approach between information users and producers to reduce barriers to technological development (Ballard, 1989). The knowledge utilization model typifies most of the initiatives in the 1980s—the "Cooperative R&D Era."

The hallmark legislation that signaled the beginning of the cooperative R&D era was the Stevenson-Wydler Technological Innovation Act of 1980. Although mostly a knowledge use initiative, it is based on components of both the dissemination and the knowledge utilization models. This legislation mandated that .5% of agency budgets be directed to technology transfer and that laboratories with budgets of $20 million must form an Office of Research and Technology Applications (ORTA) to facilitate technology transfer. It encourages personnel exchanges between federal labs, universities, and industry and allows industry to keep royalties from cooperative research agreements made with federal labs. The two most significant impacts of this act had to do with the development of ORTAs, whose functions include R&D project assessment and dissemination of information about federally owned technologies with potential commercial application, and the fact that it provided explicit directions to agencies to participate in technology transfer. Although this act has been heavily criticized, it did instigate an important change in the technology transfer mentality. It was instrumental in bringing technology transfer efforts to a level of importance that warranted a high level of public attention. The attention from the federal level was apparent over the next decade through numerous pieces of legislation.

The Bayh-Dole Act of 1980 was another significant piece of legislation in the 1980s. To encourage innovation, it allowed small businesses and nonprofit organizations to patent and retain the title to inventions that derived from federally funded R&D (De La Barre, 1986). The 1983 Presidential Memorandum on Government Patent Policy extended this to all government contractors. Neither the Bayh-Dole Act nor the Stevenson-Wydler Act, however, reflect the differing orientations and institutional environments of the laboratories involved (Bozeman et al., 1987; Bozeman and Crow, 1991).

In 1986 the Federal Technology Transfer Act (P.L. 99-502) amended the Stevenson-Wydler Act and permitted federal agencies to delegate authority to government-operated laboratories to collaborate with other agencies, industry, state and local governments, and nonprofit organizations through cooperative R&D agreements (CRDAs). The purpose of the act was not only to facilitate cooperative research but also to protect government's legitimate concerns (U.S. General Accounting Office, 1989). This act also formally established the Federal Laboratory Consortium for Technology Transfer, which falls under the dissemination model of technology transfer. Its main function is to help federal laboratories to transfer technologies. While the Technology Transfer Act has been viewed as a general success (U.S. General Accounting Office, 1989; U.S. Department of Commerce, 1989), it is not without its problems. For example, many agencies have had significant problems protecting proprietary information that resulted from a cooperative agreement; there are significant barriers to the commercialization of software; and private sector use of federal facilities is not as widespread as it should be.

Technology transfer efforts at the federal level may not appropriately be classified as "economic development initiatives." Understanding their function within the innovation process, however, helps us to better understand their relationship to economic development as well as the more economic development-oriented activities for the states. Although economic impacts are an end purpose of many of these federal initiatives, it is not a central goal of the legislation. Federal technology transfer efforts should be viewed more in terms of a response to concerns over the competitiveness crisis.

When assessing the effectiveness of technology transfer in meeting economic development or any other kinds of goals, it is important to remember the stage of the research, development, and innovation cycle that a particular initiative is targeting. Furthermore, as described in a previous section, the time line of scientific efforts is often very lengthy. As described above, federal-level initiatives tend to focus on the dissemination and use of various information and technologies. They do not address market concerns or address themselves to the marketplace. They serve an important role, however, in lowering the barriers to technology transfer, which in many cases is a premarket, or economic development, condition. Furthermore, an understanding of technology transfer efforts at the federal level are important in understanding efforts at a micro level. Technology transfer initiatives at the state level may more realistically be viewed in terms of potential contribution to economic development. These state-level efforts mirror national concerns but in a more targeted manner. The primary objective of state economic development policies, including technology-based policies, is to provide jobs (Barke, 1990)—and job creation is one of the many definitions of economic development.

The States, Technology Transfer, and Economic Development

As states jump deeper into economic development activities, they are increasingly instituting programs to attract, nurture, and develop high-technology industries (Eisinger, 1988). Of the 43 states in 1988 that had at least one program encouraging technological innovation (Minnesota Department of Trade, 1988), 26 states had a technology transfer mission as part of their technology programs. While economic research suggests the importance of technology to long-run economic growth, there is no unified body of theory to guide the states in their development efforts. Their efforts to develop new, innovative firms into sources of regional growth and (eventually) national competitiveness are, in essence, experiments in innovation and growth policy (Osborne, 1988; Eisinger, 1988). Some argue that state leaders are acting less on the basis of economic theory than on the political need to appear to be doing something (Feller, 1988).

So what have states done? Eisinger (1988) suggests that they have forged ahead without theory. He argues that "state and local development strategies typically evolve incrementally, without any underlying economic theory" (Eisinger, 1988). This argument is based on the premise that all economic development policies emerge from political pressures to "do something" (Anton, 1989). But this overstates the case. States have used basic understandings about technology's role in the economy as a basis for action; studies of the impact of R&D on these understandings have been enlarged by theories of the entrepreneurial role in high-tech firm formation and by regional economic theories that seek to explain how and why high-tech firms begin and develop (Markusen at al., 1986).

State initiatives emphasize the "upswing" of Schumpeter's (1950) growth cycle, facilitating entrepreneurial activities in an attempt to spur the next growth cycle. The rationale for technology transfer programs at the state level is that these programs facilitate the transmission of new technologies from the lab to the private sector. In turn, these technologies can become the impetus for new business creation, the introduction of new product lines to selected firms, or the revitalization of mature industries (Minnesota Department of Trade, 1988).

Technology transfer at the state level is achieved through a variety of outreach mechanisms—industrial extension services, centers of excellence, advanced technology centers—or via other government-industry-university relationships. The following provides some examples of technology transfer efforts.

Industrial Extension

Nine states currently support 13 state-supported technology extension services that provide direct consultation for technology deployment (Feller, 1991). In addition, many states have some form of an Industrial Extension Service, which often involves a statewide network of field representatives who provide technical assistance and counseling to individual manufacturing businesses.

Centers of Excellence

In 1983 New Mexico's Legislature created five Centers of Technical Excellence. These centers are associated with the state's university system and provide research services to one or more of the federal laboratories in the area. The goal of the centers is to encourage economic development through the creation of scientific and engineering facilities focusing on commercially viable R&D (Minnesota Department of Trade, 1988).

Advanced Technology Centers

Missouri's Department of Economic Development has designated four Centers for Advanced Technology on university campuses (Minnesota Department of Trade, 1988). A mission of these centers is to enhance the state economy through scientific research and technology transfer.

The New Jersey Commission on Science and Technology sponsors several programs that emphasize the role of science and technology in fostering economic development. A significant portion of New Jersey's initiatives involve Advanced Technology Centers (Minnesota Department of Trade, 1988). Each of these centers reflects the importance of technology transfer to state officials with a high level of technology transfer and educational activities (Feller, 1991). New Jersey also has Technology Extension Centers that provide education and information exchanges in cancer diagnosis, polymer processing, information services, and aquaculture (Minnesota Department of Trade, 1988).

Georgia's Advanced Technology Development Center (ATDC), affiliated with the University System of Georgia, focuses on the relationship between high-technology industries and economic development. As part of this mission, ATDC is involved in academic partnerships with universities to develop commercialization programs, provide managerial assistance, and develop other technology transfer activities (Minnesota Department of Trade, 1988).

Government-Industry-University Relations

In Illinois, Technology Commercialization Centers (TCC) were developed for location on university campuses. They are a network of technical centers that link research expertise with business and industry needs. A central mission of these centers is to enhance the transfer and commercialization of technologies from academic institutions and other research centers to private firms (Minnesota Department of Trade, 1988). The state of Delaware has formed the Delaware Research Partnership, a consortium of the state, the University of Delaware, and industry, whose purpose is to encourage cooperative research projects and the transfer of knowledge and technology between its partners. Two of the largest and most widely recognized state technology programs are Pennsylvania's Ben Franklin Partnership Program and Ohio's Edison Program. Both of these are active in technology transfer efforts. Pennsylvania's partnership program involves several targeted programs bringing together universities, industry, and government to stimulate technological innovation and business growth. This program is the primary example of a state strategy that emphasizes the targeting of selected technologies for commercialization (Feller, 1991). The state of Ohio's Technology Transfer Organization serves as a link to technical and management experts for Ohio businesses. In addition, the Edison program includes nine Edison Technology Centers, where a portion of their mandated activities relate to technology transfer.

Some state-level technology transfer initiatives are more directly influenced by federal acts. Both the 1986 Federal Technology Transfer Act and the Federal Laboratory Consortium emphasize cooperative arrangements between federal laboratories and universities, industry, and state and local government. As a result, many states are taking advantage of this relationship with the federal government to develop more targeted approaches to technology-based economic development. The catalyst in many of these relationships is some form of technology transfer. For example, the state of Tennessee is working with Oak Ridge National Laboratory to develop technological industries in the area (Barke, 1990).

State- and Federal-Level
Technology Transfer Success

Most technology transfer programs are designed to "get the science into the hands of the entrepreneurs." By doing so, government programs are attempting to jump-start entrepreneurial activity centered on new technologies, leading to a new cycle of economic growth. From our discussion of the innovation literature, it should be clear that technology alone is not enough to lead to economic growth. Government programs fail to address the market forces that affect the technological innovation and diffusion processes.

Simply getting technology into the hands of entrepreneurs will not necessarily lead to a successful product or economic benefits. Even if a successful product is developed, in technical fields firms need to continually refine the process by which they manufacture their products. Failing that, they will lose out to competitors. Too often government programs address the initial transfer of technology that leads to new products, but they do not address the process improvements that are needed by firms for them to maintain a competitive status.

There are several issues that must be addressed in assessing the success of technology transfer efforts. As discussed, the relationship between technology transfer and economic development is not straightforward. Most studies assessing the effectiveness of federal and state technology transfer efforts are only able to provide single-case or anecdotal evidence. There are few systematic or multicase data available that provide information on the successes and failures of technology transfer. In this chapter, we have proposed one framework for assessing the effectiveness of technology transfer efforts. Further wide-scale research needs to be done using a framework, such as the one presented in this chapter, to gather more useful data on the technology transfer process. Many of the programs described above have been successful. Others have not. The purpose of this chapter has been to present technology transfer initiatives together with the theoretical basis for those efforts.

In conclusion, technology transfer should not be viewed as an immediate-results economic development tool. Myriad events affect whether or not a technology results in economic growth. Many of these events lay beyond the control of policymakers. For many of the reasons cited above, technology transfer addresses one portion of the innovation and commercialization cycle. What does this mean for policies designed to nurture high-technology industry? First, policies need to address the full range of the innovation process. Simply expanding research capabilities does not ensure long-term economic growth. It should be clear from our discussion that technology transfer is not a straightforward process. Successful policies will address not just the technology but also the social process of technology transfer.

Second, simply nurturing start-ups does not ensure that diffusion will occur. Policies need to address the uncertainty prevalent in decisions to adopt an innovation. Firms need technical capacity, suitable organizational structures and processes amenable to innovative behavior, and a way to analyze and understand market signals. This is where technology transfer becomes an important process and where an appropriate role for government becomes clear. Government cannot do everything, but it can design sensible policies that address the full innovation process, not simply the supply of technology or the start-up of new firms. In terms of technology transfer, government can and should play an important role in lowering the barriers to cooperative R&D and providing the infrastructure and incentives by which technological progress and subsequent economic impacts may be achieved.

Notes

1. For discussion, see Papadakis (1990).
2. For a fuller discussion, see Kennedy and Thirlwall (1972).

References

Anton, T. (1989). "Exploring the Politics of State Economic Development Policy." *Economic Development Quarterly* 3: 339-346.

Ballard, P., T. James, et al. (1989). *Innovation Through Technical and Scientific Information: Government and Industry Cooperation*. New York: Quorum.

Barke, R. (1990). "Technology and Economic Development in the States: Continuing Experiments in Growth Management" in J. Schmandt and R. Wilson (eds.), *Growth Policy in the Age of High Technology: The Role of Regions and States*. Boston: Unwin Hyman.

Bozeman, B. and M. Crow (1991). "Technology Transfer from U.S. Government and University R&D Laboratories." *Technovation* 2(4): 231-242.

Bozeman, B. and M. Fellows (1988). "Technology Transfer at the U.S. National Laboratories: A Framework for Assessing Policy Change." *Evaluation and Program Planning* 2: 66-75.

Bozeman, B., D. Rahm, and M. Crow (1988). "Technology Transfer to Government and Industry: Who? Why? And to What Effect?" Paper presented at the Harvard Business School Colloquium on Operations and Production Management (November 16).

Burton, D. F. (1989). "Technology and Global Competition: The New Frontier for Science Policy." Testimony before the Senate Subcommittee on Science, Technology and Space (September 28).

Bush, Vannevar (1954). *Science: The Endless Frontier: A Report to the President on a Program for Postwar Scientific Research* (reissued in 1960). Washington, DC: National Science Foundation.

Cohen, S. and J. Zysman (1987). *Manufacturing Matters: The Myth of the Post-Industrial Economy*. New York: Basic Books.

De La Barre, D. M. (1986). "Federal Technology Transfer Act of 1985: PL99-502 at a Glance." *The Journal of Technology Transfer* 2 (1): 19-20.

Doctors, S. (1981). "State and Local Government Technology Transfer" in S. Doctors (ed.), *Technology Transfer by State and Local Government*. Cambridge, MA: Oelgeschlager, Gunn, & Hain.

Dosi, G., et al. (1988). *Technical Change and Economic Theory*. New York: Frances Pinter.

Eisinger, P. (1988). *The Rise of the Entrepreneurial State: State and Local Economic Development Policy in the United States*. Madison: University of Wisconsin Press.

Feller, I. (1988). "Evaluating State Advanced Technology Programs." *Evaluation Review* 12: 232-252.

Feller, I. (1991). "American State Governments as Models for National Science Policy." *Journal of Policy Analysis and Management* 11 (2): 288-309.

Fosler, R. S. (ed.) (1988). *The New Economic Role of American States: Strategies in a Competitive World Economy*. New York: Oxford University Press.

Freeman, C. (1983). *The Economics of Industrial Innovation*. Cambridge: MIT Press.

Hicks, D. (1988). "Introduction and Overview" in D. Hicks (ed.), *Is Technology Enough?* Washington, DC: American Enterprise Institute.

Kennedy, C. and A. P. Thirlwall (1972). "Surveys in Applied Economics: Technical Progress." *The Economic Journal* 82 (325): 11-72.

Kingsley, G. (1990). "Conceptualizing Technology Transfer." Unpublished working paper.

Kline, S. and N. Rosenberg (1986). "An Overview of Innovation" in R. Landau and N. Rosenberg (eds.), *The Positive Sum Strategy: Harnessing Technology for Economic Growth*. Washington, DC: National Academy Press.

Lambright, W. H. (1979). *Technology Transfer to Cities: Process of Choice at the Local Level*. Boulder, CO: Westview.

Landau, R. (1989). "Technology and Capital Formation" in D. Jorgenson and R. Landau (eds.), *Technology and Capital Formation*. Cambridge: MIT Press.

Mansfield, E. (1982). "How Economists See R&D." *Research Management* (July): 23-29.

Markusen, A., P. Hall, and A. Glasmier (1986). *High Tech America: The What, How, Where, and Why of the Sunrise Industries*. Winchester, MA: Allen and Unwin.

Minnesota Department of Trade and Economic Development, Office of Science and Technology (1988). *State Technology Programs in the United States: 1988*. St. Paul: Author.

Mowery, D. and N. Rosenberg (1989). *Technology and the Pursuit of Economic Growth*. New York: Cambridge University Press.

Osborne, D. (1988). *Laboratories of Democracy*. Boston: Harvard Business School Press.

Osborne, D. (1990). "Refining State Technology Programs." *Issues in Science and Technology* 6(4): 55-61.

Papadakis, M. (1990). *Bringing Technology to Market*. Unpublished dissertation.

Reisman, A. (1989). "Technology Transfer: A Taxonomic View." *Journal of Technology Transfer* 16(2): 38.

Schmandt, J. and R. Wilson (eds.) (1990). *Growth Policy in the Age of High Technology: The Role of Regions and States*. Boston: Unwin Hyman.

Schumpeter, J. (1950). *Capitalism, Socialism, and Democracy*. New York: Harper & Row.

Stoneman, P. (1987). "Some Aspects of the Relation Between Technological Change and Economic Performance" in *Economic Analysis of Technology Policy*. Oxford: Clarendon.

Tornatzky, L. G. and M. Fleischer (1990). *The Process of Technological Innovation*. Lexington, MA: Lexington.

Tushman, M. and W. Moore (eds.) (1982). *Readings in the Management of Innovation*. Cambridge, MA: Ballinger.

U.S. Department of Commerce (1989). *The Federal Technology Transfer Act of 1986: The First Two Years* (Report to the President and the Congress from the Secretary of Commerce). Washington, DC: Author.

U.S. General Accounting Office (1989). *Technology Transfer: Implementation Status of the Federal Technology Transfer Act of 1986*. Washington, DC: Author.

Watkins, C. and J. Wills (1986). "State Incentives to Encourage Economic Development Through Technological Innovation" in D. Gray et al. (eds.), *Technological Innovation Strategies for a New Partnership*. Amsterdam: New Holland.

Weingarten, F. (1989). "Federal Information Policy Development: The Congressional Perspective" in C. McClure and P. Hernon (eds.), *U.S. Government Information Policies: Views and Perspectives*. Norwood, NJ: Ablex.

Willinger, M. and E. Zuscovitch (1988). "Towards the Economics of Information Intensive Production Systems: The Case of Advanced Materials" in G. Dosi et al. (eds.), *Technical Change and Economic Theory*. New York: Frances Pinter.

Suggested Readings

Barke, R. (1990). "Technology and Economic Development in the States: Continuing Experiments in Growth Management" in J. Schmandt and R. Wilson (eds.), *Growth Policy in the Age of High Technology: The Role of Regions and States*. Boston: Unwin Hyman.

Bozeman, B. and M. Crow (1991). "Technology Transfer from U.S. Government and University R&D Laboratories." *Technovation* 2(4): 231-242.

Bozeman, B. and M. Fellows (1988). "Technology Transfer at the U.S. National Laboratories: A Framework for Assessing Policy Change." *Evaluation and Program Planning* 2: 65-75.

Daneke, G. (1986). "Revitalizing US Technology? A Review Essay." *Public Administration Review* (November/December).

David, P. A. (1986). "Technology Diffusion, Public Policy, and Industrial Competitiveness" in Ralph Landau and N. Rosenberg (eds.), *The Positive Sum Strategy: Harnessing Technology for Economic Growth* (pp. 373-391). Washington, DC: National Academy Press.

Dosi, G. et al. (1988). *Technical Change and Economic Theory*. New York: Frances Pinter.

Eisinger, P. (1988). *The Rise of the Entrepreneurial State: State and Local Economic Development Policy in the United States*. Madison: University of Wisconsin Press.

Feller, I. (1991). "American State Governments as Models for National Science Policy." *Journal of Policy Analysis and Management* 11 (2): 288-309.

Fosler, R. S. (ed.) (1988). *The New Economic Role of American States: Strategies in a Competitive World Economy*. New York: Oxford University Press.

Gray, D., et al. (eds.) (1986). *Technological Innovation Strategies for a New Partnership*. Amsterdam: New Holland.

McGowen, R. and E. Ottensmeyer (1989). "Symposium on Differing Perspectives on Economic Development." *Policy Studies Journal* 17 (3): 551-556.

Mowery, D. and N. Rosenberg (1989). *Technology and the Pursuit of Economic Growth*. New York: Cambridge University Press.

Osborne, D. (1988). *Laboratories of Democracy*. Boston: Harvard Business School Press.

Osborne, D. (1990). "Refining State Technology Programs." *Issues in Science and Technology* 6(4): 55-61.

Schmandt, J. and R. Wilson (eds.) (1990). *Growth Policy in the Age of High Technology: The Role of Regions and States*. Boston: Unwin Hyman.

Watkins, C. and J. Wills (1986). "State Incentives to Encourage Economic Development Through Technological Innovation" in D. Gray et al. (eds.), *Technological Innovation Strategies for a New Partnership*. Amsterdam: New Holland.

12

Theories of Entrepreneurship

TIMOTHY BATES

This chapter describes and analyzes several major strands of the vast literature on entrepreneurship that has been forthcoming in recent years. The writings of sociologists and economists are emphasized, and analyses focusing upon immigrant and minority entrepreneurs are discussed in some detail.

Three highly interrelated elements of entrepreneurship include (a) the decision to enter self-employment; (b) once self-employed, the decision to remain in business, as opposed to closing down; and (c) "success" in self-employment, measured by firm growth, profitability, and/or job creation. Some studies discussed in this chapter implicitly or explicitly encompass all three of these elements, while others focus on a specific aspect such as job creation. The more theoretical discussions of entrepreneurship tend to be all-encompassing, while the applied studies are more apt to be limited to a single element.

Qualitative studies of self-employment often distinguish "entrepreneurs" from the mere self-employed by attributing various dynamic, often intangible, traits to the former group. The self-employed are merely out to make a living while the entrepreneurs are the shakers and movers that remold the trajectory of society. The entrepreneurs "know from the start that they are trying to build a significant corporation. . . . They are driven by a desire to create an innovative force in the corporate world" (Birch, 1987: 31). The terms *entrepreneur* and *self-employed* are used interchangeably in this chapter, as they are throughout most of the social science literature. Operational, quantifiable measures of a supercharged entrepreneurial subset have been poorly developed in the entrepreneurial literature to date.

Broad Approaches to Entrepreneurship

This section begins by describing those sweeping theoretical discussions that are widely applicable to the various aspects of entrepreneurship.

Studies of a more applied nature are introduced later to evaluate and extend the concepts put forth by theorists.

Entrepreneurship as Viewed by Economists

Economists explaining the decision to pursue self-employment have traditionally emphasized the monetary rewards of entrepreneurship relative to alternative employment. An increase in the expected payoff of self-employment (holding wages constant) causes people to shift out of wage labor status in favor of entrepreneurship, and, conversely, an increase in wages (holding self-employment returns constant) induces people to abandon entrepreneurship in favor of employee status. Absent barriers that complicate small firms' formation, the choice hinges upon the opportunity costs of being self-employed as opposed to being an employee (Bradford and Osborne, 1976). Highly educated and skilled potential entrepreneurs are particularly sensitive to the opportunity costs of self-employment because business ownership often entails sacrificing high-wage positions as employees. Between 1870 and 1972 the self-employed as a proportion of the nonfarm labor force persistently declined in the United States. From the economists' viewpoint, this long-run decline reflected market signals indicating that employee status offered higher earnings potential than entrepreneurship.

More recent theoretical literature by economists portrays individuals who possess varying levels of entrepreneurial ability choosing between employee and self-employment status. Lucas (1978) suggests that persons entering self-employment are those possessing relatively more entrepreneurial ability than those choosing to be employees; ability (business acumen) is a known parameter in his model. A troubling aspect of the Lucas approach is its limited insight into causes of the huge self-employment rate differentials that exist across the different gender and racial subgroups. Do the very low self-employment rates typifying women and blacks reflect their relative lack of business acumen? Moore (1983) argues that the presence of employer discrimination in the labor market would tend to increase female and minority presence among the self-employed: Because self-employment is a method of avoiding sexist or racist employment practices that depress wage income, "blacks or women should be relatively overrepresented among the self-employed" (1983: 496). Clearly, such over-representation does not describe the labor market behavior of women or African Americans.

The traditional economist view of self-employment offers little insight into explaining why aggregate patterns of monetary rewards produce more entrepreneurs in some groups than in others. For example, in every decennial census between 1880 and 1980, the foreign born reported higher self-employment rates than native-born Americans (Light, 1972: chap. 1; Light and Sanchez, 1987). In their analysis of persons reporting a single ancestry group in the 1980 decennial census, Fratoe and Meeks (1988) found that 4.9% of all persons in the 50 largest ancestry groups were self-employed. Across the 50 groups, rates of self-employment ranged from a high of 11.7% for Russians to a low of 1.1% for Puerto Ricans. Among the predominantly black ancestry groups, self-employment rates were Sub-Saharan African, 1.4%; Dominican, 1.5%; Haitian, 1.6%; and Jamaican, 2.2%. For Chinese, Japanese, and Koreans, in contrast, self-employment rates exceeded the 4.9% average figure by wide margins, exceeding 6.0% for each of the three groups.

A small literature exploring barriers to self-employment provides some insight into the varying rates of self-employment observed across groups. In their extension of the Lucas self-employment entry model, Evans and Jovanovic (1989: 808) conclude that possession of financial capital is vitally important: "Liquidity constraints tend to exclude those with insufficient funds at their disposal." The presence of financial capital barriers to small business entry would certainly explain the lower self-employment rates observed among African Americans. Economists studying self-employment entry into manufacturing have also stressed the role played by barriers to entry. Bain (1956) argued that entry into manufacturing was deterred by capital intensity barriers, and Acs and Audretsch (1989) found that a higher capital to labor ratio was associated with a lower rate of small firm entry among 247 manufacturing four-digit SIC industries. Finally, White (1982) found that lines of manufacturing in which large financial capital investments are a prerequisite for firm formation tend to have minimal small business presence.

Entrepreneurship From the Perspective of Sociologists

Sociologists explain entrepreneurship by focusing broadly upon the social environment, which encompasses cultural as well as economic factors. Max Weber, for example, claimed that the diffusion of Puritan theology enhanced the quantity and quality of entrepreneurs in Western Europe.

Sociologists in recent years have developed theories of entrepreneurship rooted in their efforts to explain the high propensity of Asian immigrants to pursue self-employment. Emphasizing the relevance of cultural factors, the sociological analysis of small business treats firm ownership as a group phenomenon, heavily dependent upon social resources available from group support networks. The entrepreneur is seen as a member of supportive kinship, peer, and community subgroups. These networks, in turn, assist in the creation and successful operation of small business by providing social resources such as financial support, sources of labor, customers, and so forth. The social resources explanation of Asian self-employment patterns has been used to explain the low rate of black business ownership. Fratoe (1988) argues that the lower rate of self-employment among blacks may be rooted in the lower levels of social resources available from the support networks available to them.

Financial support. What precise forms do these social resources take and how do they assist business formation and operation? Sociologist Ivan Light (1972) has argued that Asians entering business have benefited from their participation in rotating credit associations. These associations commonly were set up by groups of Asian immigrants that shared some important common trait, such as former residence in the same village in China. Rotating credit association members were therefore likely to be well acquainted, even apart from their association membership. Association members each made regular cash contributions, thereby creating a pool of savings from which members could borrow for such purposes as business formation.

Protected markets. A second major source of social capital, according to Light (1972), is rooted in the loyal customers who constitute a captive, or "protected," market for minority enterprises. This form of social capital derives from the culturally based tastes of ethnic minorities that can be best served by coethnic businesses. Particularly in the early years of settlement, immigrants are assumed to patronize coethnic enterprises, and this pattern of patronage seems to typify Asian enclaves in areas such as New York City.

Sources of labor. Waldinger (1986: 159) argues that social capital in the form of loyal, low-cost coethnic employees may explain why self-employment is advantageous to Asian immigrants. "Recruiting through kin and friendship networks promoted a paternalistic relationship between immigrant owners and their employees." New arrivals often seek out employment in an immigrant firm where they can work in a familiar environment with others who know their language. Ethnicity provides a common ground on which workplace rules are negotiated. "Authority can be secured on the basis of personal loyalties and ethnic allegiance rather than on harsh discipline" (Aldrich et al., 1990: 38).

The ethnic enclave. The role that small immigrant-owned businesses play in establishing an ethnic enclave involves a coming together of the different types of social capital identified above. The number of Cuban-owned firms in Miami grew from an estimated 919 in 1967 to 8,000 in 1978 to nearly 21,000 in 1982 (Portes and Bach, 1985). Many of the early Cuban refugees in the 1960s had been business owners or professionals in pre-Castro Cuba, and these refugees often arrived in Miami with considerable financial capital. Yet the Cuban business community in the 1960s grew slowly until Cubans began to perceive their migration to South Florida as a no-return situation; this perception spurred investment in small business. According to Portes (1987), 75% of Miami's Cuban residents mostly patronized stores owned by Cubans, and 82% exclusively read Spanish language newspapers. Successive waves of migration from Cuba furnished the Miami entrepreneurs with a ready supply of low-wage workers, many of whom later established their own small businesses. Finally, Miami's Cuban community has been able to generate internally the financial capital that enabled a substantial small business community to emerge within the enclave economy.

Of the various forms of social capital associated with immigrant minority entrepreneurship, the protected market concept has been the most controversial. Available empirical evidence strongly suggests that serving a clientele of coethnics has weak causal links with the incidence or the success of immigrant-owned businesses, Asian owned or otherwise. According to Aldrich and Reiss (1976), the immigrant business that limits itself to the ethnic market sharply limits its growth potential. Fratoe reports that Asians are *much* less likely to sell to a minority clientele than are black or Latino-owned businesses (1988). Finally, empirical studies suggest that serving a captive market of coethnics is associated with an increased incidence of firm discontinuance. For example, Mexican American firms, particularly small-scale retail firms, often do cater to coethnic immigrants. This orientation appears to be an absolute hindrance to small business viability, because the very low incomes of most recent immigrants constrain the attractiveness of this protected market. A comparison of Mexican American businesses conducted by Bates and Dunham (1992) found that the firms serving a predominantly minority clientele were smaller (measured by annual sales) and more likely to go out of business than firms operating in the broader economy.

Self-employed Korean immigrants are very heavily concentrated in retailing, but the distinguishing trait of Korean business in America is their emphasis on selling nonethnic products in the general markets. Waldinger's theory (1986) of ethnic succession provides a complementary explanation of the self-employment patterns of immigrant groups such as Koreans. According to the ethnic succession approach, self-employment opportunities arise when natives opt out of certain industries because of (a) uncertain economic rewards, (b) long hours, (c) low economic returns relative to alternatives, and/or (d) low status of small business operation. Ethnic succession is most pronounced in the smaller, low-profit fields, because children do not want to succeed their parents in such firms. Thus small grocery stores in New York City have passed from Jewish to Italian to Korean ownership; wholesalers, in contrast, continue to be the domain of white ethnics. The process of ethnic succession has permitted immigrant minorities to become dominant in garment contracting fields in New York and several other large cities. Particularly noteworthy is the fact that ethnic succession is most apparent in the least attractive lines of small business.

The social resources explanations put forth by Light (1972), Light and Bonacich (1988), Waldinger (1986), Aldrich et al. (1990), and others to explain the high incidence of Asian self-employment rest, in fact, on a weak empirical foundation. The relevance of social resources in the form of loyal, low-cost coethnic employees to business viability is called into question by Fratoe's (1988) finding that small businesses owned by Asians are considerably less reliant upon minority employees than blacks and Latinos. Available evidence suggests, on balance, that the greater success of Asian Americans in pursuing self-employment may be rooted in the fact that they use the types of social resources touted by sociologists substantially *less* than their black and Latino cohorts.

Sociological studies of minority entrepreneurship have most commonly focused upon groups of recent immigrants operating firms in one specific location—Koreans in Los Angeles (Light and Bonacich, 1988), Chinese in New York (Waldinger, 1986), and Cubans in Miami (Portes and Bach, 1985). Generalization of resultant findings is risky, in part because these studies typically do not use sophisticated statistical methodology to sort out and pin down cause-and-effect relationships between individual or group traits and business success. For example, Light and Bonacich (1988) note that recent Korean immigrants effectively use family resources such as cheap labor to develop successful businesses. These Korean immigrants, however, also tend to be highly educated persons with white-collar work experience, and many of them possess substantial financial capital. It is thus difficult to assess qualitatively the relative importance of the numerous attributes of Korean entrepreneurs that are responsible for success in small business. Sociologists who discard statistical analysis as a methodological tool free themselves to speculate on various, less tangible determinants of small business success. The resultant studies tend to be provocative and interesting, yet empirical studies provide evidence that is hard to reconcile with the social-resource explanations of minority business behavior. Another case in point is the elusive rotating credit association.

Although rotating credit associations do exist, they do not appear to be a major force in financing Asian-owned businesses. In his study of New York City's garment industry, Waldinger (1986: 136) found that "the emergence of immigrant-owned garment firms owes little to the role of ethnic credit-raising associations . . . only two of the Chinese firms attributed any assistance to family associations or rotating credit associations." According to Bates and Dunham (1992), Asians report that their most common source of debt for financing small business formation is commercial bank loans; loans from family rank a distant second and loans from friends are third. Thus only 16.9% of the business start-up loans to Asians come from friends, and only a portion of this would represent the contribution of rotating credit associations. In their most recent book, Light and Bonacich (1988) report no concrete evidence of Korean immigrants using rotating credit associations.

Narrowly Focused Analyses of Entrepreneurship

Both the sociologists and the economists have recognized the importance of access to financial capital as a prerequisite for entry into most lines of self-employment. Studies of small business behavior discussed below highlight other prerequisites for entry, especially the possession of work experience, educational credentials, and specific skills (broadly referred to as "human capital"). Those lacking the requisite financial and human capital resources, quite apart from their desire to pursue entrepreneurship, are unable to enter most lines of self-employment. Those who establish small businesses in spite of their paucity of relevant skills and financial capital often find that their tenure in self-employment is short lived: they are quite likely to go out of business soon after entry. A series of studies in recent years has analyzed barriers to self-employment. These barriers are most commonly analyzed in the context of discrete stages of self-employment: (a) barriers to entry and (b) barriers that lessen the likelihood of firm survival.

A counterpart of self-employment barrier studies is a series of analyses seeking to explain why it is that highly educated and skilled Asian immigrants choose to operate marginal lines of business such as grocery stores and garment factories. In contrast to barriers that keep potential entrepreneurs from operating small businesses, "blocked mobility," it is argued, forces skilled immigrant entrepreneurs to remain self-employed despite their preference for salaried work as employees. The applicable studies of barriers and blocked mobility are discussed as three interrelated stages of entrepreneurship: (a) self-employment entry, (b) small business survival, and (c) firm growth.

Self-Employment Entry

What are the characteristics of those who choose to enter self-employment? In their time series analysis of self-employment entry among nonminority men, Evans and Leighton (1989) found a strong, positive association between greater asset holdings and self-employment entry. Further, they observed that the percentage of nonminority men who were self-employed increased with age until the early forties and then remained constant until retirement. Evans and Leighton (1989) investigated several human capital measures and found some evidence that college graduates, other things being equal, are particularly prone to pursue self-employment.

In their analysis of self-employment entry using Census Bureau Survey of Income and Program Participation (SIPP) data, Bates and Dunham (1992) found that Asians were the group most likely to enter self-employment, while blacks were least likely. A multivariable logit analysis of entry revealed, however, that the Asian characteristic, by itself, was not a statistically significant determinant of entry into self-employment. The significant determinants of self-employment entry were household wealth, owner age, gender, education, marital status, and the African American racial characteristic. Those most likely to enter self-employment possessed household wealth exceeding $100,000; they were in their late thirties or forties, married, and male; and they possessed five or more years of college education; finally, they were not African American. Asians, according to this analysis, entered self-employment in large numbers because, as a group, they were the most highly educated and they were more likely to possess household wealth of $100,000 plus—more so than nonminorities or other minorities. Blacks, in contrast, have lower rates of business entry because of their lower levels of wealth and educational attainment and they are less likely to be married. The statistical significance of the racial characteristic indicates that blacks, irrespective of wealth or education, face a different, less congenial set of business formation opportunities than other groups.

Among women only, higher education and years of work experience are strong predictors of self-employment entry (Bates, 1992). Certain industries (skill-intensive services, for example) are particularly open to women who are highly educated, while other industries (construction) appear to be male preserves.

The above studies suggest that entry into self-employment is a process shaped by the individual characteristics and resources of potential entrepreneurs, interacting with industry-specific factors; personal traits associated most strongly with entry into self-employment are wealth holdings, education, and age (a proxy for years of work experience). Groups characterized by high net assets and advanced educational credentials are overrepresented in the small business universe. Immigrant Koreans, for example, are natural candidates for self-employment entry because many emigrate with substantial financial capital; most Korean adult immigrants are college graduates, often possessing substantial managerial or professional work experience. Blocked mobility, according to Aldrich et al. (1990: 72), is an additional impetus to self-employment: "As middle-aged immigrants with poor facility in English, many Koreans find it difficult to reenter white-collar occupations; hence business has emerged as an alternative path of upward mobility." In his study of 63 Chinese-owned garment firms in New York City, Waldinger (1986) found that most of the owners felt that self-employment was not the best way to get ahead in U.S. society. Most stated a preference to work as employees for large firms but felt they lacked the appropriate linguistic and technical skills.

Small Business Survival

Jovanovic's (1982) model of small business behavior analyzes the impact of uncertainty regarding entrepreneurial ability on the growth and survival of young firms. Jovanovic assumes that entrepreneurs are highly uncertain about their managerial abilities when they enter self-employment. The behavior of young firms varies more (and produces higher failure rates) than that of older firms because the owners' estimates of their abilities are not at all precise. Owners gradually learn about their managerial abilities by running the business and observing how well they do. As their learning progresses, they acquire increasingly precise estimates of their abilities. As owners learn more about their abilities, those who revise their personal ability estimates upward tend to expand output, whereas those who revise their ability estimates downward tend to contract or to dissolve their businesses: "Efficient firms grow and survive; the inefficient decline and fail" (Jovanovic, 1982: 650).

Hypotheses suggested by Jovanovic's model were tested by Evans (1987) in his analysis of 100 manufacturing industries. Consistent with Jovanovic, Evans found that younger manufacturing firms with smaller sales volumes are the most likely to go out of business. Bates and Nucci (1989) compared small businesses started before 1976 (older firms) and those started between 1976 and 1982 (younger firms). The younger firms (a) were more likely to have discontinued business operations by 1986, (b) had smaller annual sales, and (c) were more dispersed around the mean value of sales. Controlling econometrically for firm age, they found that firms with larger sales volumes were much more likely to remain in business than smaller firms.

While a near consensus exists on the relationship between firm size, age, and likelihood of discontinuance, Birch (1987: 18) asserts that "smaller companies are actually less likely than larger ones to liquidate a facility" during the 1981-1985 time period. Birch's figures are completely incompatible with the firm survival rate data examined by Bates and Nucci over the 1982-1986 time period. The Birch analysis mixes the concept of the "establishment" with that of the firm; the latter may consist of more than one establishment. Nonetheless, the closure rates reported by Birch for the larger small businesses are so vastly higher than those reported by other studies that the Birch results are highly suspect.

On the related topic of job creation by small businesses, Birch (1987) has produced estimates of the share of net new jobs for the economy originating in the small business sector that are very substantially out of line with estimates generated by other studies. Birch's claim (1987) that "very small firms have created about 88% of all net new jobs" during the 1981-1985 period is completely inconsistent with the findings of related studies (Armington and Odle, 1982; Brown et al., 1990).

On the issue of firm size and survival among young firms, Bates (1993) found that the single strongest predictor of firm annual sales was the amount of financial capital invested in the small business. This finding relates to the earlier discussion on barriers to small business entry: Persons with significant financial resources are the most likely to enter self-employment. Larger financial investments in one's business start-up, in turn, are associated with greater sales volume, and the firms achieving the larger sales volumes are the ones that are most likely to remain in business. In an analysis of the traits of owners that are associated with firm survival, Bates (1990) found that the young firms most likely to remain in business were the ones headed by entrepreneurs having four or more years of college and who had made substantial financial capital investments (equity and debt) into their small business ventures. Thus the entrepreneurial traits associated with self-employment entry—substantial wealth and advanced educational credentials—mirror the traits associated with small business survival. Wealth and advanced education, finally, increase one's access to commercial bank financing, and this access, in turn, increases the total financial resources available for small business investment. Persons lacking financial and human capital resources, in contrast, have very little access to institutional sources of debt: They are therefore often unable to put together the financial investment needed to enter into many of the larger-scale lines of small business. If they nonetheless choose to establish very small firms, their chances of survival in business are low.

Government efforts to assist small business have often focused upon providing small amounts of debt capital to aspiring entrepreneurs, particularly minority business owners. Studies have shown that black entrepreneurs in particular are often thwarted in their business ventures by extreme financial capital scarcity (Bates, 1989). Government programs that provide loans to minorities may therefore open up opportunities for creating new firms and expanding existing ones. Yet these programs frequently target loan assistance to low-income persons who lack the education and skills that are prerequisites for success in most lines of self-employment. No serious studies of government loan programs have demonstrated that small amounts of debt can overcome human capital deficiencies that otherwise minimize chances for business success (see, for example, Bates and Bradford, 1979: chap. 9). In fact, available evidence overwhelmingly indicates that small loans, targeted to low-income individuals lacking the requisite skills for operating a business, very frequently become delinquent loans that are never repaid (Bates, 1984). Debt capital and owner human capital endowments are complements in the small business world, not substitutes.

The Small Business Administration (SBA) extended tens of thousands of Economic Opportunity Loans (EOLs) during the late 1960s and 1970s to persons whose "total family income from all sources is not sufficient for the basic needs of the family" (SBA, 1970: 6). Most EOL recipients were minorities and few possessed the necessary skills, education, or work experience that successful business operation commonly requires. Implementing a loan program whose guiding philosophy required that most recipients be bad credit risks had predictable results: Rates of loan delinquency and default were extraordinarily high, exceeding 70% for African American-owned small business start-ups (Bates and Bradford, 1979: chap. 9). Yet clones of the disastrous EOL program continue to be widespread in the 1990s at state and local government levels.

Successful public sector loan programs assisting small minority (and non-minority) businesses have targeted the higher income, better educated owner who possesses appropriate skills and experience for operating a viable small business. When assistance is provided to higher income entrepreneurs, the objection that invariably arises is this: "Why help those who are already successful?" The response is straightforward: It is the viable firms that generate economic development and create jobs. Their profits support investments that permit future expansion and job creation. The alternative—supporting nonviable firms—simply creates mass loan default and business failure, as demonstrated by the SBA's now defunct EOL program.

Firm Growth

Back in 1981 David Birch reported that small businesses were generating 80% of the new jobs created by the U.S. economy. This finding struck a responsive chord and was used throughout the 1980s as a justification for government efforts to assist the growth of small business. Serious researchers in recent years have heaped scorn upon Birch's methodology and his empirical findings (White and Osterman, 1991; Brown et al., 1990; Bates and Nucci, 1989); nonetheless, the early work of Birch—irrespective of its validity—has put the topic of small business growth and job creation very much in the policy spotlight.

Net job creation, by size of firm or establishment, is a tricky topic to analyze, and it is best approached from several perspectives. Consider the highly accurate data on Milwaukee private sector employment growth from 1979 to 1987, reported by White and Osterman (1991). Net job creation for Milwaukee is summarized below, by size of establishment:

Number of Employees	Change in Jobs, 1979-1987
1-19	14,968
20-49	6,628
50-99	9,382
100-249	16,778
250+	−40,390
Total	7,366

These data seemingly support the Birch hypothesis that net job creation is concentrated in the small business sector. Yet an analysis of these very same data, by industry sector, produces a very different conclusion:

Sector	Change in Jobs, 1979-1987
manufacturing	−51,127
services	60,534
all	7,366

From this perspective, employment changes appear to reflect a transformation of Milwaukee's economy from a base of manufacturing to one of services. Given that job creation is dominated by service firms, is it nonetheless true that small firms are creating most of the new jobs? The following figures, for services *only*, do not indicate a dominance of small firms in the job creation process:

Number of Employees	Change in Jobs, 1979-1987
1-19	10,519
20-49	5,491
50-99	9,765
100-249	16,064
250+	18,695
Total	60,534

Which firm group generated the 7,366 net new jobs in Milwaukee? Was it the establishments with fewer than 20 employees in all industry groups or was it the service firms having 250-plus employees per establishment? The former group produced 14,968 new jobs while the latter generated 18,695. Posing the question in this way is simply superficial and obscurantist. Clearly, the generation of 7,366 net new jobs in Milwaukee reflects economic base transformation, job creation by service firms of all sizes, as well as other factors such as the cyclical influence of the 1982 recession on manufacturing employment.

Abstracting from job creation, what do we know about small firm growth? Analyses of geographic areas where small business growth is particularly pronounced reveal that high rates of small business formation are key to the growth process: "Birth, not rebirth, is the hallmark of the most dynamic places in the country" (Birch, 1987: 24). Birch notes that rapidly growing areas such as Charlotte, North Carolina, often experience high business loss rates, while declining areas like New Haven, Connecticut, are characterized by low loss rates. Loss rate variations explain neither job generation patterns nor net growth of local business communities. Rather, the growth dynamic is dominated by the small business formation rate.

In a rare instance of empirical confirmation, Birch's analysis of geographic area growth is highly consistent with the findings of Stevens (1984), who investigated growth rate differences in self-employment among racial groups. In his analysis of very large samples of black, Asian, and Hispanic self-employed, based upon Census Bureau data, Stevens observed the following annual rates of entry and exit from self-employment:

Group	Entry	Exit
Black	14.9%	12.3%
Hispanic	17.2%	13.9%
Asian	20.7%	12.3%

These percentage figures mean literally that the percentage of black firms formed (discontinued) in an average year was equal to 14.9% (12.3%) of the total number of black firms existing in 1972. The Asian and black business communities exhibit equal exit rates; the faster growth of the number of Asian-owned firms is attributed entirely to their higher rate of entry. Further, Stevens found that this pattern typified growing and declining industry groups:

Group	Entry	Exit
Growing: business services	23.1%	13.1%
Declining: auto dealers	10.5%	14.0%

The obvious policy implication is that small business development depends vitally on the rate at which new businesses are formed.

Policy Implications

What useful information might a practitioner concerned with promoting small business learn from the above literature on the theory and practice of entrepreneurship and small business behavior? A consistent set of traits broadly describes those who are most likely to enter self-employment and to operate businesses that are likely to survive. The top entrepreneurial candidates most commonly command financial capital and human capital resources that are appropriate for successful operation of small businesses. Many of the successful entrepreneurial candidates are highly educated and skilled, which means that career options other than small business ownership are most likely open to them. Highly educated people tend to gravitate toward certain lines of self-employment, particularly professional services, business services, and finance, insurance, and real estate. People possessing abundant financial capital resources tend to establish the larger small businesses, and they disproportionately gravitate toward the capital-intensive fields, particularly manufacturing and wholesaling (Bates, 1992).

Thus potential entrepreneurs who make the plunge into self-employment are largely those with the resources to overcome the barriers to entry, and they are properly seen as responding to opportunities. The small business-promoting policymaker may therefore seek to lessen entry barriers through devices such as small business loan programs or subsidized credit schemes. Entrepreneurship may be made more attractive through inducements such as preferential procurement programs targeting government contracts to small businesses generally or to a small business subset such as minority-owned firms. Tax incentives, such as the ubiquitous calls for lower rates of taxation on capital gains, may improve the attractiveness of self-employment to potential high-income entrepreneurs. Yet tax incentives alone will do nothing for the potential entrepreneur whose business plans are blocked by entry barriers such as limited capital access.

Small business may also be promoted by permitting high rates of immigration, especially if the immigrants are affluent, highly educated, and experienced prior to their arrival in America. Self-employed Asians (most of whom are immigrants) pay much less attention than blacks and nonminorities to the opportunity costs of self-employment, according to Bates (1988). Because of language barriers and the like, they tend to cling to self-employment even if their own labor incomes are low and wage employment seemingly offers higher potential earnings. This relative indifference to the opportunity costs of self-employment reflects barriers such as language difficulties that limit opportunities for wage and salary employment.

In their study of Los Angeles immigrant entrepreneurship, Light and Bonacich (1988) observed heavy Korean ownership of small firms, many of which served low-income, predominantly black communities. A popular line of speculation is this: Why did local residents not take greater advantage of these business opportunities? Bonacich and Light noted that the long hours and low pay associated with operating such firms is unacceptable to most native-born workers: "Most had other job opportunities that paid better, had shorter hours, and provided greater security" (Light and Bonacich, 1988: 367). But what about the local poor, the unemployed, the welfare recipients: Why did they not seize the opportunities exploited by the Koreans? The answer, quite simply, is that few of the local poor had the college educations, white-collar work experience, or financial resources possessed by so many of the immigrant Korean entrepreneurs.

References

Acs, Z. and D. Audretsch (1989). "Small Firm Entry in Manufacturing." *Economica* 56: 255-265.

Aldrich, H., and A. Reiss (1976). "Continuities in the Study of Ecological Succession: Changes in the Race Composition of Neighborhoods and Their Businesses." *American Journal of Sociology* 81: 846-866.

Aldrich, H., R. Waldinger, and R. Ward (1990). *Ethnic Entrepreneurs*. Newbury Park, CA: Sage.

Armington, C. and M. Odle (1982). "Small Business: How Many Jobs?" *Brookings Review* 20: 14-17.

Bain, J. (1956). *Barriers to New Competition*. Cambridge, MA: Harvard University Press.

Bates, T. (1984). "A Review of the Small Business Administration's Major Loan Programs" in Paul Horivtz and Richardson Pettit (eds.), *Sources of Financing for Small Business*. Greenwich, CT: JAI.

Bates, T. (1988). *An Analysis of Income Differentials Among Self-Employed Minorities*. Los Angeles: Center for Afro-American Studies, University of California.

Bates, T. (1989). "Small Business Viability in the Urban Ghetto." *Journal of Regional Science* 29: 625-643.

Bates, T. (1990). "Entrepreneur Human Capital Inputs and Small Business Longevity." *The Review of Economics and Statistics* 72: 551-559.

Bates, T. (1992). "Self-Employment Entry Across Industry Groups." Unpublished manuscript.

Bates, T. (1993). *Banking on Black Business: The Potential of Emerging Firms for Revitalizing Urban Economies*. Washington, DC: Joint Center for Political and Economic Studies.

Bates, T. and W. Bradford (1979). *Financing Black Economic Development*. New York: Academic Press.

Bates, T. and C. Dunham (1992). "Facilitating Upward Mobility Through Small Business Ownership" in George Peterson and Wayne Vroman (eds.), *Urban Labor Markets and Individual Opportunity*. Washington, DC: Urban Institute Press.

Bates, T. and A. Nucci (1989). "Small Business Size and Rate of Discontinuance." *Journal of Small Business Management* 27: 1-8.

Birch, D. (1981). "Who Creates Jobs?" *Public Interest* 65: 3-14.

Birch, D. (1987). *Job Creation in America*. New York: Free Press.

Bradford, W. and A. Osborne (1976). "The Entrepreneurship Decision and Black Economic Development." *American Economic Review* 66: 316-319.

Brown, C., J. Hamilton, and J. Medoff (1990). *Employers Large and Small*. Cambridge, MA: Harvard University Press.

Evans, D. (1987). "The Relationship Between Firm Size, Growth, and Age: Estimates for 100 Manufacturing Industries." *The Journal of Industrial Economics* 35: 567-582.

Evans, D. and B. Jovanovic (1989). "An Estimated Model of Entrepreneurial Choice Under Liquidity Constraints." *Journal of Political Economy* 97: 808-827.

Evans, D. and L. Leighton (1989). "Some Empirical Aspects of Entrepreneurship." *American Economic Review* 79: 519-535.

Fratoe, F. (1988). "Social Capital of Black Business Owners." *Review of Black Political Economy* 16: 33-50.

Fratoe, F. and R. Meeks (1988). *Business Participation Rates and Self Employment Incomes: An Analysis of the 50 Largest Ancestry Groups*. Los Angeles: Center for Afro-American Studies, University of California.

Jovanovic, B. (1982). "Selection and the Evolution of Industry." *Econometrica* 50: 649-670.

Light, I. (1972). *Ethnic Enterprise in America*. Berkeley: University of California Press.

Light, I. and E. Bonacich (1988). *Immigrant Entrepreneurs: Koreans in Los Angeles, 1965-1982*. Berkeley: University of California Press.

Light, I. and A. Sanchez (1987). "Immigrant Entrepreneurs in 272 SMSAs." *Sociological Perspectives* 30: 373-399.

Lucas, R. (1978). "On the Size Distribution of Business Firms." *Bell Journal of Economics* 9: 508-523.

Moore, R. (1983). "Employer Discrimination: Evidence from Self-Employed Workers." *The Review of Economics and Statistics* 65: 495-501.

Portes, A. (1987). "The Social Origins of the Cuban Enclave Economy of Miami." *Sociological Perspectives* 30: 340-372.

Portes, A. and R. Bach (1985). *Latin Journey*. Berkeley: University of California Press.

Small Business Administration (SBA) (1970). *SBA: What IT is . . . What IT Does.* Washington, DC: U.S. Government Printing Office.

Stevens, R. (1984). "Measuring Minority Business Formation and Failure." *Review of Black Political Economy* 12: 71-84.

Waldinger, R. (1986). *Through the Eye of the Needle.* New York: New York University Press.

White, L. (1982). "The Determinants of the Relative Importance of Small Business." *The Review of Economics and Statistics* 54: 42-49.

White, S. and J. Osterman (1991). "Is Employment Growth Really Coming from Small Establishments?" *Economic Development Quarterly* 5: 241-257.

Suggested Readings

Aldrich, H., R. Waldinger, and R. Ward (1990). *Ethnic Entrepreneurs.* Newbury Park, CA: Sage.

Bates, T. (1984). "A Review of the Small Business Administration's Major Loan Programs" in Paul Horivtz and Richardson Pettit (eds.), *Sources of Financing for Small Business.* Greenwich, CT: JAI.

Bates, T. (1989). "Small Business Viability in the Urban Ghetto." *Journal of Regional Science* 29: 625-643.

Bates, T. (1990). "Entrepreneur Human Capital Inputs and Small Business Longevity." *The Review of Economics and Statistics* 72: 551-559.

Bates, T. and C. Dunham (1992). "Facilitating Upward Mobility Through Small Business Ownership" in George Peterson and Wayne Vroman (eds.), *Urban Labor Markets and Individual Opportunity.* Washington, DC: Urban Institute Press.

Birch, D. (1987). *Job Creation in America.* New York: Free Press.

Brown, C., J. Hamilton, and J. Medoff (1990). *Employers Large and Small.* Cambridge, MA: Harvard University Press.

Evans, D. and B. Jovanovic (1989). "An Estimated Model of Entrepreneurial Choice Under Liquidity Constraints." *Journal of Political Economy* 97: 808-827.

Evans, D. and L. Leighton (1989). "Some Empirical Aspects of Entrepreneurship." *American Economic Review* 79: 519-535.

Jovanovic, B. (1982). "Selection and the Evolution of Industry." *Econometrica* 50: 649-670.

Light, I. and E. Bonacich (1988). *Immigrant Entrepreneurs: Koreans in Los Angeles, 1965-1982.* Berkeley: University of California Press.

Stevens, R. (1984). "Measuring Minority Business Formation and Failure." *Review of Black Political Economy* 12: 71-84.

Waldinger, R. (1986). *Through the Eye of the Needle.* New York: New York University Press.

White, S. and J. Osterman (1991). "Is Employment Growth Really Coming from Small Establishments?" *Economic Development Quarterly* 5: 241-257.

PART VI

Theoretical Perspectives

13

Constituting Economic Development: A Theoretical Perspective

ROBERT A. BEAUREGARD

More so than such policy arenas as income maintenance, nuclear deterrence, and intergovernmental aid, that of economic development displaces critical assessment. Its inherent sensibility, avowed pragmatism, and unflinching optimism overwhelm intensive probing of its theoretical tendencies and ideological biases. Economic development seems like such an appropriate thing to do, regardless of one's political persuasion, that one cannot criticize without being seen as a nay-sayer and an opponent of progress.[1] Who can argue with an endeavor that attempts to improve the production and consumption of the goods and services that make our lives enjoyable and fulfilling?

Uncritical admiration, however, stifles a probing of the constitutive rules that structure our understandings of what economic development means and that delineate what actions and consequences qualify as valid and appropriate (Giddens, 1979: 48-95; Shapiro, 1981: 95-126). Constitutive rules are the seldom-considered assumptions and theoretical relationships that direct our thinking. Failure to critically evaluate them hides ideological biases and suppresses the inevitable tensions and disagreements that arise in a multiethnic society of tenacious inequalities, precarious democratic practices, and deeply ingrained capitalist values.

The purpose of this chapter is to expose the constitutive rules of economic development as practiced in the United States by local and state governments and nonprofit development corporations (Clarke and Gaile, 1992; Eisinger, 1988; Fainstein and Fainstein, 1989). My remarks are directed at the "mainstream" of economic development practice rather than at alternative governmental formulations (Fitzgerald and Cox, 1990; Giloth and Mier, 1989) or community-based economic development initiatives (Giloth, 1988). Although the latter often challenge the logic of mainstream economic development, that logic dominates public perception, political debate, and policy initiatives.

My approach involves two essential steps in any theoretical endeavor: the drawing of boundaries and the filling of categories (Zerubavel, 1991). The first divides those actions, conditions, and ideas that are meaningful or useful from those that are not. Without such a partitioning, we cannot proceed to develop theoretical arguments. The second step, the filling of categories, reflects on procedures to gather and validate knowledge. Categories need to be given substance and, although they suggest what is important, the relevance of specific information and the credibility of sources still need to be decided. Here we confront the validity and utility of knowledge. Each step thus responds to, even as it reproduces, the constitutive rules of any theoretical or practical endeavor.

My intellectual probing is meant to show how different ways of acting and knowing are privileged while others are neglected. It reveals an economic development process tightly bound to the core institutions of capitalist society, generally dismissive of peripheral activities (even those essential to capital investment), and constantly groping for a stable knowledge base with which to justify its actions. Such understandings have both ideological and practical consequences: tempering claims to the inherent sensibility of economic development, debunking the mystification intrinsic to this profoundly political act, encouraging greater precision in how we gauge its effects, and making us more sensitive to its capitalist underpinnings.

Drawing the Boundaries

There is no better place to begin than with the term *economic development,* a term that implies and elaborates a series of divisions and categories that are anchored in and reflect its constitutive rules. By "unpacking" economic development, we can identify actors and actions privileged by mainstream practice, reflect on the social structures within which these actors are embedded, and reveal conceptual attitudes toward time and space, two crucial dimensions of any theoretical perspective.

Initially, the adjective *economic* seems an innocuous modifier of *development.* Yet, it harbors profoundly important distinctions involving social action and knowledge disciplines (Zukin and DiMaggio, 1990: 39-118) and, in the context of economic development, has a specific meaning that reproduces important societal divisions and makes economic development more exclusive than commonly recognized. *Economic* does not refer simply to all things economic.

In the realm of contemporary Western thought, proclaiming something to be economic is meant to distinguish it initially from the political—political economy being a central theme of the great nineteenth-century social theorists—and secondarily from the social and cultural.[2] Economic development is about what happens when individuals and organizations engage in the production, distribution, and consumption of goods and services. It is not to be confused with the exercise of power (unless one considers the weight of economic wealth to be inherently an issue of power), self-governance and public debates over collective goals (excepting labor organizations and government involvement in economic development), the social interactions of people (ignoring, for example, social interactions at the workplace and the impact of household formation on labor force participation), and the values and traditions that define the culture (except when they affect the work ethic, propensities to save, or entrepreneurial inclinations).

This list of what is excluded from economic development points to the fictions inherent in the label "economic." One can hardly construct the list without adding caveats that weaken the clarity of the distinctions. The boundary between the economic and the noneconomic is revealed as problematic.

The economic distinction, however, is less an attempt to delineate appropriate actions than it is to identify actors who are essential to the process: those who invest capital in *specific* types of economic activity (Bartik, 1991: 3). *Economic* confers a privileged status on those who have access to capital and who use it to start, operate, and expand businesses. In practice, the distinction is even more precise. It divides those whose investments are "meaningful" for the core institutions of capital from those that are not, a point to be elaborated later. Irrelevant are actors without capital, even though they might engage directly in business activity (e.g., wage and salary workers) or provide essential infrastructural assistance (e.g., bank regulators).

Second, the modifier *economic* is an ideological statement meant to deflect attention from the inherently political nature of economic development (regardless of whether government is actually involved) and to act as a buffer (available when needed) between key investors and elected officials and government bureaucrats who might introduce the scrutiny and accountability of a democratic society. The constitutive rules of economic development policy are designed to reinforce the "capitalist trenches" (Katznelson, 1981). To say *economic* is to give a warning: Political involvement in economic development must be carefully controlled and severely limited. The government can support and subsidize economic development, but it is not a capitalist—it is not a primary actor in the economic development arena (Offe, 1975).

Now, all of this might seem fairly obvious if not analytically pedestrian. Probing a bit more into the meaning of *economic,* we uncover additional tendencies and distinctions that further narrow its meaning. Take, for example, the constant mentioning of "jobs" as the prime objective of economic development policy. One hears this repeatedly in public pronouncements and reads it over and over again in policy documents, and one might well assume that economic development really refers to the "development of employment."

In fact, the economic development process subordinates job growth to capital investment and, moreover, is frequently inattentive to the *quality* of jobs being produced. "Jobs" are ideologically and politically useful, but in actuality more attention is paid to investors and business owners. Evidence for this lies in the institutional separation of economic development from employment and training activities and the weak linkages between them (Van Horn et al., 1986). The operative theoretical statement is this: Without investors, there is no employment. In addition, the privileging of investors and business owners, rather than workers, and the political imperative of job growth biases economic development officials against careful scrutiny of the types of jobs being produced.

This argument has interesting implications. First, it ignores small-scale economic activity involving one or two individuals operating with only their "sweat equity" as investment. "Investors" are ontologically distinct from "workers" and those who are both (that is, using their own labor to create a business) are dismissed. This both undermines "indigenous" development and creates a hierarchy of actors in which workers and small business people are less important than large investors. Investing capital, not investing labor, is preferred. Second, it disallows economic activity based upon bartering or any other form of nonmarket interactions (for example, ad hoc collaboration). Third, it confines the investment process to capital that flows through commercial banks, pension funds, and venture capitalists. Ethnic or family savings associations are ignored and even community banks are not favorably considered. Fourth, it assumes that capital investment is unproblematically linked to job creation, when we know that capital invested in machines might well reduce the number of jobs and that, while simply having a job is important, the quality of the employment and the wage rate have significant consequences for economic development. Finally, it dismisses the Marxian notion that value resides in the labor process and instead puts the emphasis on the "risk" of the investor as if workers were disposable participants.

These distinctions, then, subdivide the total realm of economic activity thereby qualifying only certain actors and actions as relevant for economic development practice. For example, any activity that we might consider "informal" is discarded. Such activities are peripheral not only to the regulatory and taxing agencies of governments but, most important, are too loosely connected to the financial institutions and wealth-generating mechanisms that define who is and who is not a serious investor of capital. Regardless of how many jobs informal entities might create or even the amounts of capital actually invested in them, they are insignificant because they are not articulated with and thus under the control of either the state or the institutional mechanisms of the "formal" economy. To this extent, economic development is constituted around formal sector business activities linked to the core institutions of capitalism. All else, no matter how "economic," is irrelevant.

In addition, these distinctions exclude sociocultural activities that contribute to the production, distribution, and consumption of goods and services but that do not directly generate profits. These include the formation of cultural and familial attitudes toward entrepreneurship and the pooling of capital, the cultural reinforcement and articulation of consumption patterns, parental socialization of children to the "world of work," and the range of activities provided by women as they manage households, raise children, and generally develop the "labor force."

For all of these reasons, the "economic" of U.S. economic development is a problematic category. Although it addresses the most powerful segments of the economy, those that include most of the country's employment, it does not encompass many significant and peripheral "economic" activities that contribute to economic activity. Moreover, "economic" dismisses the importance of political involvement. To this extent, economic development is tightly bound by divisions that privilege the core institutions of capitalist society.

Of interest, economic development is only tangentially about the enhancement of the institutions that make capital investment possible and successful: banks and other financial intermediaries, education and training entities, business schools, patent offices, legal structures, social networks of entrepreneurs and inventors, mutual support groups (more common in the informal sector), and governments. Until only recently, the emphasis of U.S. economic development policy was wholly on growth—more jobs, more businesses, larger businesses, higher per capita incomes—rather than on what is commonly meant by development—the enhancement of the capacity to act and to innovate. State and local programs concerned with entrepreneurship, the technological capacity of businesses, venture capital, and education exist, but they enhance the capacities of businesses and not the capacities of institutions that support businesses. Thus, even when economic development turns to institutional development, businesses and not the structures that support businesses are the target.

Clearly, the institutional structures that frame formal sector economic activity can enhance or detract from the viability of the economy. Infrastructural entities contribute directly to employment, regional domestic production, and local business activity. Most important, they provide services that make investment and production possible. When these services (for example, venture capital or legal advice on export procedures) are lacking, the growth potential of economic activities is reduced. Their peripheral status would thus seem antithetical to the objective of enhancing growth. Nevertheless, growth remains dominant both because growth can be induced within a much shorter time frame than the developmental capacity of institutions can be changed and because short-term political considerations require growth.[3]

In fact, neither development nor growth can be understood without reference to specific notions of time. States of development and growth are arrayed along continua. Development ranges from underdevelopment to full development and growth from decline or stagnation to robust expansion. Implicit is a linear temporal order defined by its destination; the respective end states are what give these processes meaning. Whereas growth is incremental and seemingly never ending, however, development is imaginatively limited and "staged."[4] Development involves movement toward either developmental efficiency (to be technical) or unfettered and unimpeded markets (to be ideological). Once attained, the only way to transcend this "perfect" state is for another "stage" of capitalism to appear; for example, postindustrial capitalism (the liberal version) or post-Fordist capitalism (the Marxist version). Developmental time is thus lumpy and growth time continuous. Both are linear, though hardly ever turning back upon themselves.

This linearity represents progress, and progress is a major justification of economic development policy. To ignore the growth of the region or locality, or to fail to engage in development, is to "fall behind" other areas. They are making progress. Their fortunes have been elevated and people are better off. Progress means improvement and, in an ever-changing and competitive world, only the lazy or the foolish ignore economic development. This is what makes it so sensible.

This perspective on time fosters a particular attitude toward history; the past is only significant as a bench mark. It is "where we were," a state from which to progress forward and thus to be quickly discarded. Although the past might also be an impetus for new policies or renewed efforts, collective memory is unimportant. Collective memory slows growth as groups in the path of change resist so as to maintain the world as they know it. To this extent, the past is not constituted as a set of commitments that need to be respected. Rather, the past is either a burden on the development process (for example, in deindustrialized regions) or a set of assets and talents to be exploited for their future-oriented possibilities (for example, in a historical theme park).

Finally, the "time" of economic development has political and economic dimensions. First are the political cycles within which policy shifts occur, for example, when a new governor reorients state policy to reflect her growth objectives and political agenda. Added to this are program funding cycles that affect the repertoire of economic development tools and the timing of outputs. Second, there are economic cycles: the business cycle as regionally or locally experienced with its impact on the "take-up" or success of economic development programs and secular trends (that is, the long cycles of development) that shape what general types of policy responses (for example, fiscal stimulus or business subsidy) will be most effective.

Economic development officials are aware of these temporal elements, but they seldom use them as part of their strategic thinking. Almost no recognition is given to the rhythms of everyday life as they are played out in the transformations of communities, the needs of workers, and the effect of aging on entrepreneurs. In addition, minimum attention is paid to the rhythms of investment and disinvestment as they occur differently in different industries, firms, and areas of the region, and across private, nonprofit, and governmental activities. Rather, time is approached in a simplistic but not wholly rationalized fashion, just as the economy is simplified but not wholly tamed.

A similar conclusion applies to economic development's spatial distinctions. Most obvious is that economic development practice operates uncomfortably within a political partitioning of economic space, a disjuncture widely recognized by practitioners and theorists. Wholly unreconciled in the United States is the tension between the spatial diffusion of economic activity and the geopolitical specificity of the goals and interventions of economic development policy. Expansive markets are incompatible with political territoriality. As a result, interventions are highly inefficient; beneficial and deleterious consequences leak into other jurisdictions.

A second spatial distinction stems from the basic confinement of the economic to capital investment that emanates from core institutions: the separation of production spaces from spaces of reproduction and, to a lesser extent, spaces of consumption (Zukin, 1991). Practitioners are little concerned with neighborhoods or public spaces except when they stand in the way of new investment or, as in the case of waterfront districts, when they have the potential to be commodified. On the other hand, retail and commercial areas are frequently prime components of local economic development strategies. In essence, the landscape is segmented and prioritized into landscapes of production, commodified consumption, and social reproduction with the symbiosis among them falling outside the boundaries of economic development.

A third spatial distinction is that between the local and the global. On the one hand, theorists and practitioners recognize the ways in which local economies are penetrated by international economic forces, and this insight has engendered a number of globally oriented activities ranging from export policies and the pursuit of foreign investment and international events (e.g., the Olympics) to the creation of "international" booster initiatives. These global pretensions, however, are constrained by the ever-present stricture that the benefits of economic development should be spatially confined. As economic development policy attempts to reconstruct the global economy within the locality or region, then, it becomes either highly inefficient or contradictory.

Finally, dominating this spatial perspective is the notion of space as relative and inert. Economic performance is gauged both in relation to the past and in comparison with other localities, regions, or nations, for example, asking how well Buffalo compares with Cleveland or the United States with Japan. The landscape is arranged comparatively and competitively, but with only a weak understanding of the perniciousness of capitalist uneven spatial development. Economic development policy in the United States is devoid of a dialectic and critical understanding of the seesawing of capital across the landscape and its devaluing of some areas *in order to* value others (Smith, 1984). Yet, practitioners and policymakers are well aware, and more and more disdainful of, the spatial competition that ensues when states or localities bid for new plants or corporate headquarters (Goodman, 1979; Milward and Newman, 1989). Their inability to opt out of such bidding wars is attributable to politics, however, rather than the dynamics of uneven development.

More theoretically, space too often functions as a container. The dominant impression is of an absolute space that is inert and nonreactive rather than historically contingent and socially created. Recognition is given to the importance of the spatial juxtaposition of activities (e.g., the locational proximity of universities and high-tech firms) but analysis stops short of viewing space as dynamic, manipulated, and reactive.

In sum, economic development involves a quite sharp partitioning of reality. Within the United States, its theoretical distinctions focus on the dominant economic institutions of capital, privilege investors, include but subordinate the state, emphasize growth over institutional capacity, offer a linear notion of time mediated by political and economic cycles, and fail to resolve a variety of spatial contradictions between political territoriality and economic space, local intervention and global influences, and land-scapes of production and those of consumption and reproduction. Absent from mainstream economic development are indigenous and informal economic activities, the needs and desires of workers, the democratizing influences of the state, a sense of developmental capacities, collective memory, and an understanding of space as socially negotiated and dialec-tically uneven.

These distinctions define the knowledge base of economic development and simultaneously constitute the rules that constrain alternative and oppositional theoretical formulations. To adopt economic development's dominant perspective is to become a prisoner of categories that are only reluctantly made explicit.

Filling the Categories

How we gather knowledge, thus filling categories, and what types of knowledge (for example, numerical data, personal beliefs, secondhand descriptions) are termed valid or useful are considerations typically sub-sumed under the rubric of epistemology, the theory of knowledge or study of how we know what we know (Sayer, 1984: 26). Traditional epistemo-logical investigations tend to be idealistic; they assume epistemology to be an intellectual practice independent of who does what in what context (Kaplan, 1964: 20-23). As materialists point out, however, one cannot separate valid knowledge and procedures from actors who are deemed legitimate carriers of that knowledge or conceive of ideas and information as anything but socially embedded.

The impression received from economic development documents is that economic development is rationalized on the basis of quantitative data that fit within a lawlike (or nomothetic) framework governing the workings of the economy (Harvey, 1969: 50-53). The key actors thus become the research-ers and technical experts who conduct surveys, gather data from secondary sources, and produce the resultant analyses (Walton and Kraushaar, 1990; Wolman and Voytek, 1990).

Most such analyses quantify the number of jobs and businesses in an area, consider how each has changed over time within specific occupa-tional and industrial categories, and relate this to changes in per capita or household income, unemployment, capacity use, and other measures of economic performance. These are the important divisions of economic reality and they are cast within a systemic framework of relationships whose objective is economic growth. In that framework, strong correlations exist between the demand for locally produced goods and services, local capital investment, job generation, household income, and the fiscal viability of governments (Eisinger, 1988: 34-54; Thompson, 1965); progress along each of these dimensions leads to economic development.

Although economic development documents are dominated by this empirical framework and by technical experts and their scientific procedures, policymakers view this knowledge as mostly valid but not always useful. Knowing the trends in capital investment by industry, for example, only crudely targets the specific actions that might enhance capital investment and produce economic growth.

Another type of knowledge thus takes priority in policy formulation. This knowledge is idiographic: particularistic stories about the economic successes of individual firms or regions and future investment opportunities (Harvey, 1969: 50-54). A plant is touted for its application of a new technology, a region's rapid growth is attributed to innovative investors, an instance of management-union cooperation is praised. Reinforcing these success stories are their opposites, instances where firms failed to improve technology, stagnant regions lacking venture capital or research universities, management-labor conflict, and businesses inattentive to export opportunities.

Correspondingly, economic development officials dedicate much of their time and effort to marketing the area, selling it and the programs they have available (Levy, 1990). To do so, they tell stories and search for "generative metaphors" that will capture the imaginations of investors, elected officials, and the public (Mier and Fitzgerald, 1991). Not empirical analyses but rhetorical strategies become essential; knowledge is politically vacuous until it is situated in a framework of meanings that motivate investors to act.

Civic boosters, investors, developers, policymakers, elected officials, practitioners, and academics all tell such stories. They speak not of trends but of possibilities. They map the paths of success and failure. From a strict epistemological standpoint grounded in an objectivist social science, the stories lack validity (Bernstein, 1983). Too idiosyncratic for researchers—that is, not fitting into a lawlike theoretical frame—they are nonetheless very powerful for they give meaning to conditions and proposed actions (Edelman, 1988).

The power of these stories derives, in part, from the actors who articulate them. Economic development policy is dependent upon investors without whom the whole process loses substance (Rubin, 1988). So privileged, the stories of investors take on an aura that exceeds their technical validity.

Investor knowledge is not only useful—valid in a policy rather than scientific sense—but also a justification for economic development itself. It is investors, industry representatives, and corporate executives who are believed to have privileged insights into potential investments and, whether right or wrong about what needs to be done, nothing will be done without them. To generate investment, then, economic development policymakers agree to listen to and seriously consider their views. If economic development is about the core institutions of capital, the representatives of those institutions hold the key to successful practice.

Stories by economic development officials from other regions and localities are also given credence. Despite the competitive nature of economic development, those with success stories are willing to share them. Successful policymakers and practitioners search for recognition and praise, and those pursuing success reach out to find examples of interventions that might be replicated in their locales.

Obviously, policymakers and practitioners do not believe that economic development is a zero sum game even though agencies still engage in competitive recruitment, debates rage over international trade policy, and investors often play one locality against another (Goodman, 1979; Thurow, 1980). Those involved in economic development seem to believe that their successes cannot be replicated or that they are so far ahead of other regions and localities that they cannot be surpassed. Sharing relevant knowledge, then, has no significant costs.

While investors, corporate executives, business representatives, economic development officials, and, less so, researchers are listened to and their knowledge considered useful, the voices of other actors are suppressed. Elected officials are not viewed as having special insights into economic development. Although one must pay attention to them, what they say is primarily important for establishing a good business climate and for developing funding and publicity. One would not, however, rely on elected officials to decide whether to make a venture capital investment in a fledgling biotechnology firm.

Workers are also excluded as a source of knowledge. They supposedly know little about investment potential, marketing opportunities, or technological innovations. The knowledge they have is too task specific and generally irrelevant for the larger issues of what to invest in and when.[5] Like the public at large, unless they are well organized, they are frequently informed but seldom consulted (Grant, 1990).

Thus what knowledge is declared valid, relevant, or useful depends upon which actors voice that knowledge. All of this suggests interesting conceptual problems involving the justification and rationalization of economic development decisions.

First, economic development policymakers can hardly avoid the discrepancies between the objectivist and lawlike knowledge and advice produced by researchers and the subjectivist and particularistic knowledge and advice provided by investors and successful practitioners. Public policy in the United States is highly rationalized and tends to favor objectivist analyses (Benveniste, 1977; Friedmann, 1987). Yet it is also political, and listening carefully to investors is highly valued. The problem is that investor stories are easily cast as political and labeled as the "serving of special interests" in contrast to the adherence to the common good as revealed through systematic analyses.

A number of responses to this dilemma are possible. One is to use objectivist perspectives to identify overall problems and general directions, but not to select specific actions. For this, policymakers would turn to subjective advice. Given that policy is intrinsically normative and that one must "go beyond the data" to decide what "should" be done, this is a defensible position.

Another response is to treat the problem ideologically. Economic development officials in the United States, particularly during the conservative ascendancy of the last two decades, have elevated successful investors, developers (more so in the 1980s than 1990s), and corporate executives to near cult status. American elites, of course, have always portrayed business people as having superior insights into public affairs. Even in the Progressive Era when reformers railed against monopolies and corporate evils, good government became equated with good business practices. Who better to give direction to economic development policy than those who have brought about economic growth and who risk their wealth in job-generating investments? Such people have learned the ways of the market, known success, and made a personal commitment. By this argument, idiographic knowledge is elevated beyond the particularistic and made to seem universal. At the same time, the argument excludes non-investors, particularly workers who do not risk anything.

A last solution is to find a place for idiosyncratic knowledge in the framework of the nomothetic. Objectivist research can be mobilized to justify actions precipitated by other means. Various techniques exist for testing the generality of single cases or fitting particular instances into larger trends.

By doing so, single opportunities are made into targets for public policy. Public policy assumes that similar interventions can be made in numerous settings. The larger the spatial scale or number of governmental units involved, in fact, the more prominent this trait becomes. Economic development policy would be something quite different if it involved only unique actions. Rather, policy is supposed to be successful through repetition across a multitude of essentially similar cases.

What is interesting about economic development practice is that it manages to combine repetition with novelty. For example, a program to subsidize investment in new technology in multiple manufacturing firms will exist in the same organization with support for a single instance of waterfront development. To this extent, economic development officials must manage the political and administrative integration of seemingly opposed epistemologies.

A second conceptual problem involves the problematic linkage between knowledge and action. This has two dimensions: first, the gap between an objectivist understanding and the normative dimension of policy and, second, the necessity to propose not what is needed or what should be done but what will be done by investors who are not subject to government command.

All public policy suffers from a chasm between the objectivist knowledge base upon which it is to be formulated and the normative dimensions of the policy environment. Public policy is not what objectively is signaled by an analysis of data and experiences. Rather, it is a projection of what we *should* do given the analyses of technical experts, policymakers' sense of the future, a reading of the political climate, and sensitivity to the values held by the community (Wildavsky, 1979). Normative choices are inherent to the policy process and to political action.

Objectivist epistemology offers no solution to the problem of how to pass from empirical findings to political action. Declining employment in a regional industry, for example, calls for either reindustrialization or deindustrialization depending upon the values and interests one brings to the decision. Given the ahistoric nature of economic development policy, deindustrialization (except where corporate elites are in positions of public influence) is the usual response (Beauregard et al., 1992). The problem is thus easily dismissed in practice, for economic development policymaking (like all policymaking) is not wholly constrained by objectivist procedures. Rather, it uses them selectively to shape and support positions and interests whose origins are not rationalist in a traditional scientific sense.

The second issue is how to solve the knowledge-action puzzle and do so in a way that generates action by those over whom one has minimal influence. This is an issue of interests and incentives, not of adequate knowledge and rational analysis. Action does not flow from the advice of experts but from the motivations and inclinations of investors and the pressures that can be put upon them. The solution here seems to be to employ multifaceted development strategies that mix targeted initiatives based upon systemic analyses with idiosyncratic initiatives based upon the inclinations of specific large investors. The combination enables policymakers to incorporate differing knowledge bases, be sensitive to political considerations, and assure a continual level of action.

In sum, economic development involves a complex web of coexisting and conflicting knowledge bases. Practice vacillates between objectivist and subjectivist epistemologies, with the contradictions often resolved in favor of privileged actors. The whole process, moreover, is greatly influenced by the need for action and a corresponding reliance on actors whose decisions are only partly motivated by analytical knowledge, resistant to government command, and tenuously linked to the spatial arenas in which policymakers operate. As a result, the path between knowledge and action is murky and discontinuous. The response is ideological formulations, the partitioning of economic development into routine activities and special projects, and a tolerance for epistemological conflict.

Conclusions

By probing the categories of U.S. economic development and the epistemology used to fill them, we learn what is and what is not important for "mainstream" practice. Our intellectual journey reveals an activity tightly bound to the core institutions of capital and sensitive to, even dependent upon, those who control the flow of investment. Nevertheless, a good portion of what one might consider economic activity and many important but ostensibly peripheral activities are excluded.

This analysis, however, is only a beginning. The objective has been to raise consciousness about the constitutive rules of economic development to facilitate a more critical understanding. Whether that critical understanding leads us to alternative formulations or to oppose the whole endeavor of mainstream economic development remains to be seen. What is clear is that we must reflect on what economic development means if we are to overcome the spurious sensibility of its promises.

Notes

1. This is not to claim that critics do not exist, only that they focus on the obvious actions, goals, and outcomes of economic development rather than on underlying assumptions and theoretical tendencies. For example, compare the quite different perspectives of Bartik (1991), Kirby (1985), Rubin and Zorn (1985), and Street (1981), though none penetrates the logic of economic development.

2. These divisions are reflected in the disciplinary boundaries that have become such a staple of U.S. academic thought and that had their origins in the rapid spread of capitalism during the nineteenth century (Wallerstein, 1991).

3. The need to have viable institutional structures is clear in the case of developing countries and countries now in transition from centrally planned to open market economies (e.g., Poland).

4. Certainly there are limits to growth, but for most economic development officials, those limits seem so distant that they need not be considered.

5. One could conceive of economic development policy that took an opposite tack, which is one of the lessons of Japanese management practices.

References

Bartik, T. J. (1991). *Who Benefits from State and Local Economic Development Policies?* Kalamazoo, MI: W. E. Upjohn Institute.

Beauregard, R. A., P. Lawless, and S. Deitrick (1992). "Collaborative Strategies for Reindustrialization: Sheffield and Pittsburgh." *Economic Development Quarterly* 6: 418-430.

Benveniste, G. (1977). *The Politics of Expertise*. San Francisco: Boyd & Fraser.

Bernstein, R. J. (1983). *Beyond Objectivism and Relativism*. Philadelphia: University of Pennsylvania Press.

Clarke, S. E. and G. L. Gaile (1992). "Postfederal Local Economic Development Strategies." *Economic Development Quarterly* 6: 187-198.

Edelman, M. (1988). *Constructing the Political Spectacle*. Chicago: University of Chicago Press.

Eisinger, P. K. (1988). *The Rise of the Entrepreneurial State*. Madison: University of Wisconsin Press.

Fainstein, S. S. and N. Fainstein (1989). "The Ambivalent State: Economic Development Policy in the U.S. Federal System Under the Reagan Administration." *Urban Affairs Quarterly* 25: 41-62.

Fitzgerald, J. and K. R. Cox (1990). "Urban Economic Development Strategies in the USA." *Local Economy* 4: 279-289.

Friedmann, J. (1987). *Planning in the Public Domain*. Princeton, NJ: Princeton University Press.

Giddens, A. (1979). *Central Problems in Social Theory*. Berkeley: University of California Press.

Giloth, R. P. (1988). "Community Economic Development." *Economic Development Quarterly* 2: 343-350.

Giloth, R. P. and R. Mier (1989). "Spatial Change and Social Justice: Alternative Economic Development in Chicago" in R. Beauregard (ed.), *Economic Restructuring and Political Response* (pp. 181-208). Newbury Park, CA: Sage.

Goodman, R. (1979). *The Last Entrepreneurs*. New York: Simon & Schuster.

Grant, J. A. (1990). "Making Policy Choices: Local Government and Economic Development." *Urban Affairs Quarterly* 26: 148-169.

Harvey, D. (1969). *Explanation in Geography*. London: Edward Arnold.

Kaplan, A. (1964). *The Conduct of Inquiry*. New York: Chandler.

Katznelson, I. (1981). *Capitalist Trenches*. Chicago: University of Chicago Press.

Kirby, A. (1985). "Nine Fallacies of Local Economic Change." *Urban Affairs Quarterly* 21: 207-220.

Levy, J. M. (1990). "What Local Economic Developers Actually Do." *Journal of the American Planning Association* 56: 153-160.

Mier, R. and J. Fitzgerald (1991). "Managing Economic Development." *Economic Development Quarterly* 5: 268-279.

Milward, H. B. and H. H. Newman (1989). "State Incentive Packages and the Industrial Location Decision." *Economic Development Quarterly* 3: 203-222.

Offe, K. (1975). "The Theory of the Capitalist State and the Problem of Policy Formation" in L. N. Lindberg (ed.), *Stress and Contradiction in Modern Capitalism* (pp. 125-144). Lexington, MA: Lexington.

Rubin, B. M. and C. K. Zorn (1985). "Sensible State and Local Economic Development." *Public Administration Review* 45: 333-339.

Rubin, H. (1988). "Shoot Anything That Flies; Claim Anything That Falls." *Economic Development Quarterly* 2: 236-251.

Sayer, A. (1984). *Method in Social Science*. London: Hutchinson.

Shapiro, M. T. (1981). *Language and Political Understanding*. New Haven, CT: Yale University Press.

Smith, N. (1984). *Uneven Development*. Oxford: Basil Blackwell.

Street, W. (1981). "Better Late Than . . . the Myths of Economic Development." *Plant Closings Bulletin* 3: 8-12.

Thompson, W. R. (1965). *A Preface to Urban Economics*. Baltimore, MD: Johns Hopkins University Press.

Thurow, L. (1980). *The Zero-Sum Society*. New York: Basic Books.

Van Horn, C., et al. (1986). "Local Economic Development and Job Targeting" in E. M. Bergman (ed.), *Local Economies in Transition* (pp. 226-247). Durham, NC: Duke University Press.

Wallerstein, I. (1991). *Unthinking Social Science*. Cambridge: Polity.

Walton, M. and R. A. Kraushaar (1990). "Ideas and Information: The Changing Role of States in Economic Development." *Economic Development Quarterly* 4: 276-286.

Wildavsky, A. (1979). *Speaking Truth to Power*. Boston: Little, Brown.

Wolman, H. and K. Voytek (1990). "State Government as a Consultant for Local Economic Development." *Economic Development Quarterly* 4: 211-220.

Zerubavel, E. (1991). *The Fine Line: Making Distinctions in Everyday Life*. New York: Free Press.

Zukin, S. (1991). *Landscapes of Power*. Berkeley: University of California Press.

Zukin, S. and P. DiMaggio (eds.) (1990). *Structures of Capital*. Cambridge: Cambridge University Press.

Suggested Readings

Bernstein, R. J. (1983). *Beyond Objectivism and Relativism*. Philadelphia: University of Pennsylvania Press.

Edelman, M. (1988). *Constructing the Political Spectacle*. Chicago: University of Chicago Press.

Eisinger, P. K. (1988). *The Rise of the Entrepreneurial State*. Madison: University of Wisconsin Press.

Harvey, D. (1989). *The Condition of Postmodernity*. Oxford: Basil Blackwell.

Katznelson, I. (1981). *Capitalist Trenches*. Chicago: University of Chicago Press.

Kirby, A. (1985). "Nine Fallacies of Local Economic Change." *Urban Affairs Quarterly* 21: 207-220.

McCloskey, D. N. (1990). *If You're So Smart: The Narrative of Economic Expertise*. Chicago: University of Chicago Press.

Sayer, A. (1984). *Method in Social Science*. London: Hutchinson.

Wallerstein, I. (1991). *Unthinking Social Science*. Cambridge, MA: Polity.

Wildavsky, A. (1979). *Speaking Truth to Power*. Boston: Little, Brown.

Zerubavel, E. (1991). *The Fine Line: Making Distinctions in Everyday Life*. New York: Free Press.

Zukin, S. and P. DiMaggio (eds.) (1990). *Structures of Capital*. Cambridge: Cambridge University Press.

14

Metaphors of Economic Development

ROBERT MIER
RICHARD D. BINGHAM

We ambitiously began this book with a goal of articulating a synthetic theory of local economic development to be useful to practitioners. We anticipated integrating various theories of explanation emerging from the social sciences and focusing them on theories of action from the policy sciences. We no sooner arrived at a chapter outline than we recognized the heights (and folly) of our ambitions. It was not just the complexity of the theoretical terrains of explanation and action that sobered us, it also was the daunting challenge of using theory to make a better professional practice.

As we observed in the Preface, the chasm is large between the language of scholarship, with its precision, boundedness, and objectivity, and that of practice, with its fluidity, instability, and subjectivity (Forester, 1989: 16). This creates a tension that often troubles social scientists. This tension is sharply cast by Donald Schön (1983: 42):

> In the varied topography of professional practice, there is a high hard ground where practitioners can make effective use of research-based theory and technique, and there is a swampy lowland where situations are confusing "messes" incapable of technical solution. The difficulty is that the problems of the high ground, however great their technical interest, are often relatively unimportant to clients and the larger society, while in the swamp are the problems of greatest human concern.

For the local development agent in Pittsburgh, discussed by Elaine Sharp and Michael Bath, or the ones in Bonaparte, Iowa, interviewed by

AUTHORS' NOTE: The genesis of the ideas in this chapter evolved from lengthy discussions over a decade between Mier and Robert Giloth. They were further developed in Mier and Fitzgerald (1991). Portions of this chapter directly draw on that work and were contributed to by Joan Fitzgerald. Finally, Bob Beauregard reviewed it and made a number of insightful and encouraging suggestions.

Marie Howland, the tasks of advancing a local economic development project are developing consensus, forging broader alliances, fighting rear guard actions, and finding a confidence-building and inspirational public language. As Mier and Fitzgerald (1991) have observed in reviewing a number of recent practitioner-oriented books on local economic development, the use of either explanatory or predictive theory is limited in this "swampy lowland."

This limitation is well illustrated by the treatment of "partnership," alluded to one way or another in many of the chapters. It is virtually impossible to read a practical description of local economic development without encountering "public-private partnerships" or "community partnerships." Yet, none of our chapters treated "partnership" as a theoretical category, firmly bounded and standardized. This should not be surprising because, rather than a scientific category, "partnership" often is a fuzzy concept whose very ambiguity is practically and administratively useful. It connotes, for example, inspirational images such as a community's leadership standing shoulder to shoulder at a ground breaking, and images like this create the possibility for strategies such as codevelopment. But the images also have multiple meanings, ones that often are contradictory and paradoxical.

The premise of this concluding chapter is that the theories of economic development presented in the preceding chapters must be viewed within frameworks like "partnership" that force us to engage with the multiple meanings of development rather than pretending they don't exist. Drawing on the work of Marris (1987), Morgan (1986), Schön (1979, 1983), and Lakoff and Johnson (1980), Mier and Fitzgerald (1991) say generative metaphors are essential to this task of confronting ambiguity and thus a key to the integration of scholarship and professional practice because they liberate the imagination and engender new understandings of problems and approaches to their solution.[1]

Schön (1979) argues that generative metaphors are both frames of reference and processes for bringing new perspectives into existence. They are ways of seeing one thing as another and, in so doing, enabling the redefinition or resetting of a problem. Morgan (1986: 12-13) elaborates:

> Metaphor is often regarded as a device for embellishing discourse, but its significance is much greater than this. For the use of metaphor implies a way of thinking and a way of seeing. . . . We use metaphor whenever we attempt to understand one element of experience in terms of another. Thus, metaphor proceeds through implicit or explicit assertions that A is (or is like) B.

Thus the "partnerships" metaphor elicits the shoulder-to-shoulder image. Yet, it also can spark very different images for business and community leaders—different pictures of who are standing shoulder to shoulder. These differing images raise basic questions not only about who stands in the front at the ground breaking and whether the development is occurring in the right neighborhood but more fundamentally about whether it should even be built.

Beauregard addresses the ubiquitousness of such conflicting understandings when he observes that "economic development involves a complex web of coexisting and conflicting knowledge bases. . . . As a result, the path between knowledge and action is murky and discontinuous." Differing images exist because generative metaphors are fluid and open ended, and their imagery is refined within specific contexts. Further, they also are molded by a set of tendencies and biases that professionals (and scholars) bring to their practice.

Schön (1983: 270) identifies the tendencies and biases as "appreciative systems," "overarching theories," "role frames," and "media, language, and repertoires." Following the pioneering work of Thomas Kuhn, Marris (1987: 119, 138) calls these interpretative systems a paradigm or

> a set of assumptions about the scope, method, and purpose of (practice and analysis) which determines its questions, the nature of its evidence, and its principles of interpretation . . . articulating the relationship (in policy) between social purpose, government and training as (in science) between question, method, and finding.

Thus the connection between theory in local economic development and professional practice lies in the interpretation of the relevance of theory in a particular context. This interpretation, says Beauregard, must reflect awareness of the paradigms that shape the scholarship. In addition, Mier and Fitzgerald (1991) argue, for the interpretation to play a constitutive and inspirational role, it must respect the metaphors, with their attendant media, language, and repertoires, that mold narration of local economic development practice and encourage alternative actions.

In light of this, we now see that the task before us, in synthesizing this book, lies not in the terrain of meta-theory, but in that of metaphor. Morgan (1986: 12) observes:

> New insights often arise as one reads a situation from "new angles," and that a wide and varied reading can create a wide and varied range of action possibilities. . . . Our theories and explanations of organizational life are based on metaphors . . . that imply *way(s) of thinking* and *way(s) of seeing* that pervade how we understand our world generally.

Thus the question of whether the theories of economic development like those reviewed in this book can contribute to building a better practice must be measured not by their powers of explanation and prediction but by their usefulness as tools of reading, thinking, seeing, inspiring, and acting.

We will attempt to consider theories of local economic development as they contribute to the articulation of metaphor and to consider metaphor as a tool of reading. As we show on Table 14.1, our chapter authors have discussed over 50 theories of local economic development from the perspective of the various social sciences they represent.

The problem of making sense of them is reminiscent of the children's fable, "The Blind Men and the Elephant":

> Four blind men found an elephant and did not know what it was.
> "It is like a log," said one, who had flung his arms around the elephant's leg.
> "No! it is like a rope," said another, who had caught hold of its tail.
> "It is more like a fan," said the third. He was feeling the shape of the elephant's ear.
> "It is something with no beginning and no end," said the fourth, who was walking round and round the animal, feeling its sides. . . .
> "What is it?" they said.

As we reflect on the theories of local economic development presented in the earlier chapters, we find a number of them clustering. Our clusters are not the usual groupings of theory into categories that define academic disciplines but into metaphors, like "log," "rope," and "fan," that we think better frame concrete development situations. The alternative frames, we hope, will invite seeing how different aspects of development can coexist in complementary and even contradictory ways.

In the balance of this chapter, we propose seven metaphors of local economic development practice and attempt to situate many of the theories we have been reading within these metaphors. You may find this effort a bit untidy—there is not, nor should there be, a one-to-one correspondence between theory and metaphor. Further, you should expect the boundaries surrounding and dividing the metaphors to be fuzzy because they are not discrete categories but generative images. After reviewing the metaphors, we will conclude with a discussion of their application to a better practice of local economic development.

Metaphor 1: Economic Development as Problem Solving

The essence of scientific and social scientific problem solving is a type of simplification that often characterizes local economic development

TABLE 14.1. Theories and Models Appearing in This Book

Chapter 1: Blair and Premus
    Transportation Cost Model
    Costs of Production Models
    Inertia
    Coevolutionary Development
    Agglomeration Economies

Chapter 2: Nelson
    Models of Development From Above
        Price Equilibrium Models
        Dynamic Disequilibrium
        Political Economy Model
    Models of Development From Below
        Territorial Development
        Functional Development
        Agropolitan Development

Chapter 3: Howland
    Central Place Theory
    Product Cycle Theory
    Profit Cycle Theory
    Neoclassical Theory
    Circular and Cumulative Causation
    Demand-Side Development
    Supply-Side Development

Chapter 4: Wiewel, Teitz, and Giloth
    Market-Based Development
    Marxian Critiques
    Political Processes
    Sociological Theories of Community
    Integrated Economic and Sociopolitical Theory

Chapter 5: Goldsmith and Randolph
    Black Capitalism
    Poverty Culture and the Underclass
    Dispersal Theory

Chapter 6: Fitzgerald
    Skills Mismatch
    Training Theory
    Education Theory
    Equity-Oriented Education and Training

Chapter 7: Goldstein and Luger
    Neoclassical Economic Theories
    Stage/Cycle/Wave Theories
    Organization of Production Theories
    Propulsive, Innovative, and Creative Theories

TABLE 14.1. Continued

Chapter 8: Holupka and Shlay
  Political Economic Theory
  Ecological Model
  Growth Machine Model

Chapter 9: Betancur and Gills
  Community Development
  Affirmative Action
  Separatism

Chapter 10: Sharp and Bath
  Psychosocial Theories
  Rational Calculus Theories
  Institutionalist Theories
  Combinatorial Theory

Chapter 11: Melkers, Bugler, and Bozeman
  Innovation Theories
  Technology Change Theory
  Technology Transfer Theories

Chapter 12: Bates
  Utility Maximization Theories
  Social Network Theories
  Barrier Theories
  Asset Base Theory

practice. The inability of the human mind to integrate or deal with the vast number of inputs to economic development often leads to an approach to problem solving built on bounded rationality.

Herbert Simon (1976; see also March and Simon, 1958; Cyert and March, 1963) argued that organizations, like people, usually act on the basis of incomplete information about various alternative courses of action and their consequences and are able to explore only a limited number of alternatives relating to a given decision. Thus, at best, they can achieve only limited forms of rationality. He concludes that organizations and individuals settle for a form of bounded rationality called "satisficing" where decisions are simply adequate and not optimal. They are based on limited information and molded by rules of thumb.

Both the understanding and the practice of economic development suffer from the imperatives of bounded rationality. The now dated but historically important transportation cost minimization models presented by Blair and Premus in Chapter 1 are excellent examples. Bounded rationality thrives on formulas and models that invite us to discard what we can't measure, don't know, or don't understand. Thus transportation cost minimization models allow one to compute optimal locations by simplification.

Similarly, many economic developers today embrace the concept of targeting industries, another clear example of bounded rationality. Economic developers frequently use techniques such as location quotients or shift-share analysis to identify industries, which, for one reason or another, seem to thrive in a given geographic location. These industries are then "targeted" for special economic development efforts. Using these types of quantitative techniques, a recent study of Chicago's economy suggested that the following industries be targeted: chemicals, steel-related industries, printing and publishing, food processing, and furniture and fixtures, to name a few (Center for Urban Economic Development, 1989).

Bounded rationality shapes the way development needs and opportunities are read and potential actions seen in three ways: in problem definition, in the processes of problem solving, and in the order it imposes on the solutions. Problem definition, the first method of shaping, occurs in a number of ways and often is indicative of a paradigmatic framework. Beauregard argues, in discussing a mainstream view of development, that bounded rationality molds our processes of gathering knowledge and our values on what "types" of knowledge are useful to economic development practice. While seemingly emphasizing the uniqueness of each situation, little time is spent exploring variability in the ways problems are seen and set.

A common approach to bringing order to the setting of problems lies in establishing distinctive geographic or social perimeters. Nelson's introduction of the broad field of territorial development and Goldsmith and Randolph's ghetto development are two geographic examples, and Betancur and Gills's separatism a social one.

The second method of shaping is process definition. Beauregard also posits a strong interrelation between the mainstream view and the tendency to rely on the tenets of positivism and technical rationality.[2] After the problem is defined, the emphasis shifts to problem solving and the selection of rational means. In this step, there often is a pattern of uniform blindness to an aesthetic dimension to local development planning that invites passion. By "aesthetic," we do not mean beauty and grandeur, which are considered to be apolitical and within the realm of realism. Rather, following White (1987), we refer to a problem-solving process open to passionate judgments—anger, outrage, empathy. These are considered value laden and highly subjective, often stretch the bounds of credibility, and frequently open the possibility of inspirational solutions.

Finally, bounded rationality imposes order on the prospective sets of problem solutions. As Schön (1979) suggested over 15 years ago, it often leads to a reframing of the development problem such that certain institutional responses are more appropriate. Common institutional frameworks include the marketplace and public-private partnerships. The key characteristic of the reframing process, according to Schön, is minimizing conflict. What strikingly underscores this from our earlier chapters is the relative paucity of conflict-based theories of development.[3] Betancur and Gills, and secondarily Goldsmith and Randolph, see conflict as central to the understanding of minority community development. Additionally, Holupka and Shlay see it as central to the functioning of the urban political economy.

In many respects, the most useful aspect of reading a development situation through the lens of problem solving is the order it imposes, bringing action within reach. From this order, bounded rationality invites the application of a variety of the theories reviewed in this book and the derivative techniques of analysis. In its artificial tidiness and a tendency toward premature closure lie the appeal of the metaphor of local development as problem solving. We believe it reaches its fullest expression in imagining local economic development in business terms, our next metaphor.

Metaphor 2: Economic Development as Running a Business

Many of our authors invite attention to what Beauregard calls the mainstream view of local economic development practice and what Marris (1987: 128-133) refers to as a corporatist paradigm. Its essential features are a view of government's role as secondary to that of business and the economy, "regulating social expectations in accordance with the requirements of the economy." Yet it also invites seeing government trying to be businesslike and, in so doing, acting more entrepreneurial in its efforts to achieve economic growth. The view draws more on organizational than on economic doctrines:

> The primary task of government is to make the most of the land, labor, skills, raw materials, infrastructures and social amenities within its jurisdiction. The intelligence of government, therefore, is likely to depend above all on how well it understands corporate behaviors, the thinking and information on which it is based. (Marris, 1987: 130-131)

Fleshing out this corporatist view is the central object of Beauregard's effort, where he attempts both to draw its boundaries and to fill in its categories of inquiry. Examples of his mainstream perspective appear in several of our chapters, for example, Nelson's "development from above" and both Holupka and Shlay's as well as Sharp and Bath's discussions of political economy theory. In addition, they were critically addressed in Wiewel, Teitz, and Giloth's market-based theories and Fitzgerald's education and training theories.

Beauregard suggests that the mainstream, or corporatist, view emphasizes progress as the transcendence of the past, the preeminence of capital investment generating growth as the vehicle of that transcendence, and the privileging of the institutions and individuals possessing capital. He also suggests that it is a static view: It largely holds constant most social and economic institutions as well as social mores, and it relegates to the distant future any confrontation with resource limitations. Not surprising, then, we find Nelson saying "mass investment in opportunities generating fewer returns than other opportunities will ultimately leave society worse off; indeed, the purpose of making capital increasingly mobile is to ultimately upgrade the welfare of world society."

In sum, a reading of a development situation through the metaphorical lens of the metaphor of running a business would see the importance of capital investment and growth, the sanctity of private markets, the value of elite partnerships with government in a supportive role, the necessity of overcoming barriers to progress, and the beauty of dispassionate, "professional" problem-solving approaches. These, of course, are indicative of the parsimony of bounded rationality.

Metaphor 3: Economic Development as Building a Growth Machine

Systems analysis has emerged as the major vehicle for the design of new technology products and production processes. As such, it inspires similar dreams in the economic development arena. Although systems analysis has evolved side by side with bounded rationality, in many respects it represents an effort to transcend artificial boundaries by demonstrating the possibility of modeling complexity. Yet, when practically applied, it confronts information limits and has much in common with bounded rationality.

The systems approach invites a look at the individual components of an allocative system and emphasizes the interconnections between those components. This analytical approach to imposing order makes systems analysis highly compatible with empirical research. The framework provides a useful terrain for the more complex statistical and mathematical techniques in common use today such as input-output analysis and econometric modeling. Yet it is reductionist in the same sense as bounded rationality in that data limits inevitably force it to simplify complex phenomenon. The conceptual 250-sector input-output model in practice becomes a 40-sector one.

In certain respects, systems analysis is like building a complex "black box." However elaborately constructed, the box serves the function of converting resources into outputs. Presumably, economic development is the system that converts resources like land, labor, and capital into gross state product. The systems approach to economic development is most clearly illustrated by Melkers, Bugler, and Bozeman's discussion of technology transfer policies. Their dissemination and knowledge use models are systems based. And their state-level programs—industrial extension centers of excellence, advanced technology centers, and so on—are the outputs of a system. Howland's circular and cumulative causation theories and Goldstein and Luger's theories of propulsive, innovative, and creative regions are other examples of applying systems theory to economic development.

Because systems analysis is allocative, it is also political. How are land, labor, and capital allocated to "optimize" growth of gross state product? A number of the chapters in this book illustrate economic development as a political system. The dominant theories supporting this concept are those focusing on the functioning of the political economy in general and the "growth machine" in particular.

Recall that the growth machine is a coalition of local government, some unions,[4] and place-bound business interests like utilities, newspapers, real estate developers, and retail establishments pursuing their agenda of downtown renewal and the creation of safe middle-income enclaves in central cities. The growth machine is thus the black box of the systems approach—a complex, interconnected web that is largely taken for granted by practitioners at least partially because they often are part of it.

Attempting to penetrate the "black box" of the relationship between the political process and the process of urban growth is the object of many of the theories in this book. It is central to the analysis of local political economy by Sharp and Bath and especially by Holupka and Shlay. It is also an important element of Wiewel, Teitz, and Giloth's Marxian critique of community development and Betancur and Gills's separatism and community development models.

It is within the "black box" that systems theory must be examined critically. Often, that box contains the unstated, implicit assumptions about the nature of the problem most economic developers are attacking. For example, a frequent bias emanating from a systems approach is a belief in the limited effectiveness of government because of the fragmentation of public services (Schön, 1979). Systems analysis tolerates no loose ends! The obvious solution, one within the grasp of systems designers, is "coordination."

The recommendation of systemic coordination uncritically dominates and mystifies public policy discourse. It ignores obvious contradictions. What is seen and labeled as "fragmentation" among public programs is glorified as "competition" among private ones. And what is coordination for some is a stifling paternalism for others. This lack of critical reflection illustrates how easily true complexity is ignored and replaced by superficial order.

In sum, the systems approach in practice becomes another form of bounded rationality, a possibly more complex one, that in almost contradictory ways both focuses attention on and mystifies the heart of the development process, the "black box" of the political economy. Its appeal is in acknowledging complexity, then seemingly bringing some order to it. For example, it is in the nature of the growth machine to define the terms of development by setting the problems to be solved, delimiting participation, directing the processes of solution, and cloaking all of it in a facade of complexity demanding ordered coordination. Systems theory has its roots in the study of natural systems, and like those fields of inquiry in their more simplistic form, tends to take for granted the problem-setting process. With the next metaphor, one that also builds on natural systems theory, we shift the focus to problem setting.

Metaphor 4: Economic Development as Preserving Nature and Place

A metaphor evoking images of nature and place provides a way to read economic development practice in sharp contrast to the dominant corporatist metaphor.[5] It is called by Marris (1987: 133-138) an ecological paradigm and by Friedmann and Weaver (1979: 89-113) a territorial one. We see elements of it in Blair and Premus's coevolutionary development, Nelson's "development from below" and his agropolitan development, Bates's social network theories, and Holupka and Shlay's ecological model.

Organismic metaphors of development have roots in the concept of natural resource limits and attendant romantic views of nature. But they also are influenced by those experiences of community action and development that see a convergence in a place between social and natural communities. The essential features of such metaphors are concurrently a demystification of growth and a respect for the right of each community of people to a "familiar" habitat. Together, these entail a recognition that development planning cannot be decided apart from a particular context. The ecological paradigm emphasizes:

Social responsibility against economic autonomy; decentralized, democratic control against remote, concentrated, corporate hierarchies of control; and understanding the whole against the abstraction of partial relationships. (Marris, 1987: 137)

Within the framework of this metaphor, with its "Johnny Appleseed" emphasis on localism, there is a rejection of the role of local development as reactive to external economic stimuli and more of an emphasis on local catalytic actions. It also invites looking beyond dominant institutional actors to nonprofit organizations and loosely tied social networks as agents for change. Finally, from the perspective of the cyclicality of nature, it sees a much tighter interconnectedness between the past and the future than the more linear view we have previously examined.

Specifying an epistemology compatible with the naturalistic metaphor is problematic. Central to the metaphor is skepticism about a scientifically based approach to understanding and practicing local development. As a type of advanced systems theory, it demands that social analysis be directed toward seeing complex patterns of rapidly changing interrelationships rather than documenting and isolating unique factors or combinations of factors constituting the nodes of the interrelations. But the human element of these systems is what is most important—it is human relationships that count the most.

Thus the naturalistic metaphor invites seeing and valuing the passion present in the economic development process—the pathos of the disadvantaged and their determination in struggling for justice, the mortification of government leaders as their actions are called into doubt, the insecurity of social scientists as their wise musings are challenged, and the anger of business and civic elites as their vision of progress is derailed. It demands, to use Schön's imagery, descending to the "swampy lowlands," where uncertainty, instability, uniqueness, and value conflict reign.

But the naturalistic metaphor does more than reject many of the tenets of the corporatist metaphor. Despite the seductive power of its images and its emphasis on man and environment, it can become somewhat rigid in its treatment of human agency, the objects of our next several metaphors. Because of its emphasis on locality, it easily can lead to a quest to preserve the status quo and thus a rejection of emphasis on social justice and equity, particularly as equity questions cross territorial or social boundaries. Thus, not surprising, it often is a metaphor expropriated by the privileged to oppose growth or racial integration in their residential enclaves. Also, with its emphasis on local interrelatedness, it easily invites minimizing the social and economic ties with other localities, regions, and nations.

Nonetheless, its contributions are substantial. A reading of a development situation through the lens of the naturalistic, "homegrown," metaphor would see the importance of local initiative, the balance between growth and distribution, the value of public intervention, the necessity of broad-based partnerships with "community" in a preeminent role, the necessity of overcoming barriers to participation, and the beauty of passionate public debate about problem-solving approaches.

Metaphor 5: Economic Development as Releasing Human Potential

A recent article asserts:

Neglect of human resources is the major competitive failing of the United States. Indeed, if one considers the sum total of the citizenry's entrepreneurial spirit, talents, skills, education, training, and their willingness to innovate cooperate, learn, and solve problems—then the key factor in the nation's economic development is clearly our people. Yet, human resources are the most underutilized (and worst managed) factor of production in the United States. In most U.S. corporations, the talents and productivity of employees are frustrated by poor organization and the unwillingness of managers to listen to their workers. As a result, organizations tap only a small percentage of the productive and innovative capacities of their employees. Similarly, state, local, and national policies fail to develop the full potential of the citizenry—education, training, and welfare institutions all fall far short of the mark. (O'Toole and Simmons, 1989: 275)

Thus it is no wonder that releasing human potential as a metaphor for economic development has become much more pervasive in the literature and suggests a more focused role for government in the development process. The metaphor is built up from a foundation emphasizing that an economy is really its people: "Above and beyond these factors is the overarching *meta-resource of human ingenuity*. It is, in fact, the dynamism of a people . . . that leads a nation to economic success" (O'Toole and Simmons, 1989: 276).

The metaphor of economic development as human resource development is found in the pronouncements of public officials, often supported by empirical research. For example, Ohio's Governor George Voinovich (1992) recently complained that Ohio businesses spend more than $240 million annually to provide basic remedial training for their workers. More than 22.5% of all incoming Ohio college freshmen, he observed, need remedial training in mathematics and 19.5% require it in English. Only

41% of Ohio's high school graduates go on to college compared with 48% nationally.

Research like this abounds in public policy debates on such issues as the level and quality of support for public education. It also demands a perspective that is not so place bound. For example, Donald and Jean Haurin (1987) found that immigration is highly cyclical and that, in a downturn relative to the rest of the nation, a state can lose a substantial portion of its youth—the next generation of the state's labor supply. Furthermore, they found that the youths migrating from depressed regions tend to be more highly educated males. This sort of research has forced the policy debate in particular states to be more global in nature.

Many of the chapters in this volume illustrate economic development as human resource development. Fitzgerald's chapter on labor force, education, and work is a rich and obvious example, but others are not so obvious. In discussing neighborhood development, Wiewel, Teitz, and Giloth discuss entrepreneurship; employment, training, and placement; and labor-based development. Similarly, Goldsmith and Randolph trace the historical debate within the African American community over integrationist versus separatist strategies as well as the importance of education and entrepreneurship to each. Also, Betancur and Gills examine the individualistic approaches of affirmative action policy. Finally, Bates, observing huge entrepreneurship differentials across racial subgroups and noting that immigrant entrepreneurs tend to be highly educated, suggests that human resource development might also be an inducement to entrepreneurship.

Compared with the preceding metaphors, the human resources metaphor rivets attention on people as the agents of social change. It also abstracts place and group as important components of development favoring instead larger systems defined by market forces. A major consequence of this, as Betancur and Gills persuasively argue and Fitzgerald amplifies, is the presence of nationality-specific groups with a high proportion and geographic concentration of disadvantaged people within a society of individual achievement and advancement. This seemingly inherent societal unjustness is the focus of our last two metaphors.

Metaphor 6: Economic Development as Exerting Leadership[6]

Throughout the literature on economic development, a dominant image is that of leadership. Although none of our chapters directly focused on this dimension of development, in one form or another it is a strong undercurrent. Sharp and Bath recognize its presence in their combinatorial theory and when they discuss the success of Mayor Richard Caliguiri in Pittsburgh. The metaphor also appears in Bates's emphasis on entrepreneurial ability and in Wiewel, Teitz, and Giloth in their discussion of the political process.

Beauregard discusses the role played by the leadership metaphor as providing the subject of success stories by and about "civic boosters, investors, developers, policymakers, elected officials, practitioners, and academics." The result, he says, has been the elevation of "successful investors, developers . . ., and corporate executives to near cult status."

Two images of local economic development leadership are prevalent: groups of elites who promote local growth, charity, or respond to social crises and individuals who run roughshod over bureaucratic impediments, whether embodied in bank redlining or urban redevelopment plans (Committee for Economic Development, 1982; Krumholz and Forester, 1990; Alinsky, 1971). The leadership structure often is a civic committee, applying tried-and-true "businesslike" approaches. The unifying and generative metaphor of exerting leadership is amplified by "partnerships" and "entrepreneurship."

Giloth and Mier (1993) argue that these traditional forms of leadership are inherently limiting in three respects. First, they generally represent a structural and formalistic response to what often is an idiosyncratic problem. Second, they often concentrate on a monumental project as the engine for development. Finally, they narrowly focus on the production of wealth, ignoring redistributive and social justice dimensions of urban problems: They are leadership without regard to the concerns of the followers (Krumholz, 1984; Judd and Parkinson, 1990). Because of these limitations, a third type of leadership is emerging in many communities.

This new leadership is derived from collaborative efforts to create solutions to difficult and controversial problems, ones with seemingly intractable conflicts of group interests. Wiewel, Teitz, and Giloth discuss it as sociological theories of community; Betancur and Gills address it with their community development model; and Sharp and Bath introduce it in their combinatorial model. In these discussions, the metaphor of "cooperation" replaces both "partnership" and "entrepreneurship" and connotes bringing diverse interests to the table to engage in "social invention" (Whyte, 1982; Osborne, 1989; Mier and Fitzgerald, 1991).

Giloth and Mier (1993) call this new leadership "cooperative leadership." They observe such collaborations occurring in task forces, ad hoc networks, innovative organizations, and partnerships borne of struggle. Betancur and Gills give meaning to this in their discussion of the world's fair debate in Chicago. Other impressive examples are when workers and owners, bankers and neighborhoods, or local governments and community coalitions jointly create solutions to commonly perceived and experienced problems (Fitzgerald and Simmons, 1991).

Wiewel, Teitz, and Giloth discuss recent sociological thinking about community decision making that underscores the importance of inter-organizational linkages or loosely coupled social networks that connect resource-mobilizing individuals. How widespread, representative, and interconnected community leadership is affects a community's capacity to engage in collaborative problem solving. As Betancur and Gills emphasize, deep divisions in community life, such as race or class, may limit the potential for cooperative leadership in the short run. Such networks may become a long-term community goal.

Sharp and Bath's discussion of civic leadership raises fundamental questions about democracy and social change. Some critics argue that citizenship, not leadership, is today's real challenge (Bradford, 1985; Giloth, 1985). A discussion of civic leadership also underscores concern that a "professionalized" leadership may undermine grass-roots action (Bookchin, 1987). Giloth and Mier (1993) address these debates without resolving them. They acknowledge that the conception of cooperative leadership, if unreflective or misapplied, may become co-optive, gloss over inequities and institutional responsibilities, or paralyze grass-roots empowerment.

Metaphor 7: Development as a
Quest for Social Justice

In the past two decades, it is the infrequent electoral campaign—federal, state, or local—that is not built around jobs and economic development. Public opinion polls consistently reveal a high public concern that the fruits of progress are failing to reach large segments of people. This is the dominant reason that so many social movements of recent years, like those for civil rights and women's rights, speak to issues of economic justice and equality. In the eyes of many people, public participation in local economic development is nothing less than a quest.

The imagery of quest is important because much of the social equity-oriented approaches to economic development are considered an aberration. Beauregard cautions us of this:

> Absent from mainstream economic development are indigenous and informal economic activities, the needs and desires of workers, the democratizing influences of the state, a sense of developmental capacities, collective memory, and an understanding of space as socially negotiated and dialectically uneven.

The themes of social justice and equality run throughout our chapters. Sometimes, their presence is by omission, as when Howland warns that "existing growth theories say little about those in poverty and consistently left behind." Other times, it is quite direct, as when Goldsmith and Randolph say "racial discrimination still plays an enormous negative role in the social experience of nearly every black American," which Bates amplifies with discussion of many barriers to minority entry to entrepreneurship. Ultimately, many of the authors argue, the greatest likelihood for successful equity-oriented local development activity hinges on the climate for group-based and community-based action.

Sharp and Bath advise us that "conflict over economic development is likely to be more sustained and more difficult to transform into collaboration in relatively prosperous communities." A case in point is the once proposed 1992 Chicago world's fair. Holupka and Shlay describe the proponents of the fair as a "classic manifestation of an urban growth machine" yet invite attention to other, more community-based, groups who ultimately shaped the public understanding of the costs and benefits of the proposed fair. Betancur and Gills also dissect the fair, focusing on the tensions and coalition building within the minority communities of Chicago. They invite attention to how the rhetoric of justice can easily play into an individualistic strategy like affirmative action and ultimately lead to less equality for the most oppressed groups of people.

Fitzgerald argues that an equity-directed, community-based, collaboration-oriented approach to local development is possible and effective. She cites labor-community collaborations in Springfield, Massachusetts, and Chicago as promising examples. Betancur and Gills reinforce her with the caveat that ultimately the success of a community development approach hinges on the issue of the degree of community control of the development process. They also observe that this control issue faces strong opposition from advocates of dispersal policies, an observation underscored by Goldsmith and Randolph. Finally, Goldsmith and Randolph, Betancur and Gills, and Bates emphasize the complex ways race enters the local development calculus.

Wiewel, Teitz, and Giloth go a step farther and argue that a general theory of community economic development is possible, and they sketch its form. They see its essential purpose as communitarian, directed toward economic growth focused on particular populations and communities, yet operating within the realities of the market. In addition, they argue that community economic development must have a strong sociopolitical dimension capable of mobilizing political power. Finally, they emphasize that "a notion of community goals retains a strong hold on the imaginations and perceptions of people who participate in community development."

The metaphor of quest for social justice complements and amplifies several earlier metaphors. It shares a strong emphasis on localism and passion with the naturalistic metaphor, a focus on people with the human resources one, and a predilection to collaboration with the leadership one. To all, it adds a dimension of social justice as fairness: a need to consider the circumstances of the least advantaged and a respect for their right to ultimately control the shape of their own futures. It forces seeing the practice of local economic development as being unconstrained by professional or bureaucratic dictates. Finally, and possibly most important, it forces confrontation with the divisive role race plays in development.

Conclusion

Forester (1989: 64) reminds us that practical work is context dependent and that contexts are enormously variable and constantly shifting. Thus, he says, prediction is hazardous and theory must take on a different role: "It is to direct the attention of the decision-maker, to suggest what important and significant actors and events and signals to be alert to, to look for, to take as tips or warnings." In local economic development, a discipline dedicated to building the future, it has an added responsibility—inspiring confidence in pathways ahead. It is our belief that the use of metaphor bridges the rich stream of theory presented in our chapters and makes theory more useful in understanding and shaping practice.

Schön (1983) describes good professional practice as a process of iteration in which a problem is framed, explored, reframed, reexplored, and so on. We hope, from this discussion, that we have showed how metaphors can provoke visualization of those frameworks. Morgan (1986) carries Schön's prescription a step further and challenges both the analyst and the practitioner to employ multiple frames *at once*. Again using metaphor as the guide to articulating frames, he advocates a process of understanding that consists of successively framing the problem and accumulating insights, framing it a second way and accumulating more insights, and so on. Then, in a manner similar to Bell (1992), Throgmorton (1992), and others, he suggests that making sense of the insights is a process of using them to shape and tell a story.[7]

In the end, it is the story that enables the practical "reading" of the development situation, one that then enhances the ability to "see" appropriate solutions. Additionally, it is easier to build political consensus for the solutions because stories are everyday vocabulary, inspire enthusiasm, passion, and commitment, and serve as an integrative vocabulary. In sum, stories derived this way are enormously practical—they translate the epistemology of social science research into the phenomenology of professional practice.

Much of the practice of local economic development is the promotion of alternative, better futures. Delgado (1989: 2416) argues that alternative visions require a new social construction of reality—new patterns of perception embedded in narrative habits and patterns of seeing. Further, he argues, we only "participate in creating what we see in the very act of describing it." So, inspirational story telling, as Beauregard also argues, takes priority in economic development policy and strategy formulation. Metaphors provide the inspirational spine to the stories.

After reading the chapters on theory and practice, we are certain that one could imagine other possible metaphorical ways to group them. We are not advocating these as the best, just ones that are faithful to the theories presented earlier. Our hope, simply, is that the reader can see this metaphorical analysis as a way of making sense of complex, paradoxical situations by bringing diverse perspectives to bear on them. We hope, to paraphrase Schön, that we have encouraged some wallowing in the swamp of ambiguity, contradiction, and paradox.

Notes

1. Two points require elaboration. We recognize that rhetorical devices, or tropes, other than metaphor contribute to understanding and interpretation. We are focusing, however, on the constitutive, or building, dimension of the practice of local economic development—thus our emphasis on the role of metaphor, among the most inspirational of rhetorics.

2. The scattering of other conflict-based theories include the Marxian critiques of community development presented by Wiewel, Teitz, and Giloth and the political economic theories of Sharp and Bath.

3. See also Schön (1983) and White (1987: 58-82).

4. The building and construction trades, in particular, and sometimes public employees, unions are active members of the growth coalition.

5. This conception follows Morgan's (1986) use of the metaphor "organization as an organism" and also draws on the work of Guba (1985).

6. This section draws heavily on Giloth and Mier (1993).

7. Other rhetorical devices such as metonymy, synecdoche, and irony serve as important elements of persuasive argument, but our emphasis has been on the inspirational role of metaphor. For a good review of story telling in planning, see Throgmorton (1992).

References

Alinsky, Saul (1971). *Rules for Radicals*. New York: Vintage.
Bell, Derrick (1992). *Faces at the Bottom of the Well: The Permanence of Racism*. New York: Basic Books.

Bookchin, Murray (1987). *The Rise of Urbanization and the Decline of Citizenship.* San Francisco: Sierra Club Books.

Bradford, Calvin (1985). "Neighborhood Reinvestment: The Legacy and the Challenge." Paper presented at the National Neighborhood Coalition Conference (November 25).

Center for Urban Economic Development (1989). *Economic Audit of Chicago: An Identification of Target Industries.* Chicago: Center for Urban Economic Development, University of Illinois.

Committee for Economic Development (1982). *Public-Private Partnerships: An Opportunity for Urban Communities.* Washington, DC: Author.

Cyert, R. M. and J. G. March (1963). *A Behavioral Theory of the Firm.* Englewood Cliffs, NJ: Prentice-Hall.

Delgado, Richard (1989). "Storytelling for Oppositionists and Others: A Plea for Narrative." *Michigan Law Review* 87 (August): 2383-2441.

Fitzgerald, Joan and Louise Simmons (1991). "From Consumption to Production: Labor Participation in Grassroots Movements in Pittsburgh and Hartford." *Urban Affairs Quarterly* 26 (4): 512-531.

Forester, John (1989). *Planning in the Face of Power.* Berkeley: University of California Press.

Friedmann, John and Clyde Weaver (1979). *Territory and Function: The Evolution of Regional Planning.* Berkeley: University of California Press.

Giloth, Robert (1985). "Organizing for Neighborhood Development." *Social Policy* 15 (Winter): 37-42.

Giloth, Robert and Robert Mier (1993). "Cooperative Leadership for Community Problem Solving" in Robert Mier et al. *Social Justice and Local Development Policy* (pp. 165-181). Newbury Park CA: Sage.

Guba, Egon (1985). "The Context of Emergent Paradigm Research" in *Organizational Theory and Inquiry: The Paradigm Revolution.* Beverly Hills, CA: Sage.

Haurin, Donald R. and R. Jean Haurin (1987). "The Migration of Youth and the Business Cycle: 1978 to 1984." *Economic Development Quarterly* 1 (May): 162-169.

Judd, Dennis and Michael Parkinson (1990). "Patterns of Leadership" in Dennis Judd and Michael Parkinson (eds.), *Leadership and Urban Regeneration: Cities in North America and Europe* (pp. 295-307). Beverly Hills, CA: Sage.

Krumholz, Norman (1984). "Recovery: An Alternative View" in Paul R. Porter and David Sweet (eds.), *Rebuilding America's Cities* (pp. 173-191). New Brunswick, NJ: Rutgers University Press.

Krumholz, Norman and John Forester (1990). *Making Equity Planning Work.* Philadelphia: Temple University Press.

Lakoff, G. and M. Johnson (1980). *Metaphors We Live By.* Chicago: University of Chicago Press.

March, J. G. and H. A. Simon (1958). *A Behavioral Theory of the Firm.* Englewood Cliffs, NJ: Prentice-Hall.

Marris, Peter (1987). *Meaning and Action: Community Planning and Conceptions of Change.* London: Routledge & Kegan Paul.

Mier, Robert and Joan Fitzgerald (1991). "Managing Economic Development." *Economic Development Quarterly* 5 (August): 268-279.

Morgan, Gareth (1986). *Images of Organization.* Newbury Park, CA: Sage.

Osborne, David (1989). *Laboratories of Democracy.* Boston: Harvard Business School Press.

O'Toole, J. and J. Simmons (1989). "Developing the Wealth of the Nation: A Call for a National Human Resources Policy." *Economic Development Quarterly* 3 (November): 275-282.

Schön, Donald (1979). "Generative Metaphor: A Perspective on Problem-Setting in Social Policy" in Andrew Ortony (ed.), *Metaphor and Thought* (pp. 254-283). New York: Cambridge University Press.

Schön, Donald (1983). *The Reflective Practitioner: How Professionals Think in Action*. New York: Basic Books.

Simon, H. A. (1976). *Administrative Behavior*. New York: Macmillan.

Throgmorton, James (1992). "Planning as Persuasive Storytelling About the Future: Negotiating an Electric Power Rate Settlement in Illinois." *Journal of Planning Education and Research* 12 (Fall): 17-31.

Voinovich, George (1992). "Cuts Won't Gut Higher Education." *Cleveland Plain Dealer* (July 26): 1-C, 4-C.

White, Hayden (1987). *The Content of the Form: Narrative Discourse in Historical Representation*. Baltimore: Johns Hopkins University Press.

Whyte, William Foote (1982). "Social Inventions for Solving Human Problems." *American Sociological Review* 47 (February): 1-13.

Name Index

Subject Index

About the Authors

Timothy Bates is Director of the Urban Policy Analysis Graduate Program at the New School for Social Research and Professor of Urban Policy Analysis. He was awarded the Ph.D. degree in economics by the University of Wisconsin in 1972. He is the author of four books on urban economic development and minority business development issues, the latest of which is *Banking on Black Business: The Potential of Emerging Firms for Revitalizing Urban Economies* (1993). His papers have been published in various journals, including the *Journal of Regional Science, Urban Affairs Quarterly, Journal of Urban Affairs, Journal of Urban Economics, Review of Economics and Statistics, Journal of Finance, Journal of Human Resources, Review of Black Political Economy, Journal of Business*, and *Challenge*. In 1988 he was the American Statistical Association/National Science Foundation research fellow at the U.S. Bureau of the Census, where he developed the small firm data file from the Characteristics of Business Owners data base.

Michael G. Bath completed a B.A. at Benedictine College, an M.A. at the University of Kansas, and is currently a doctoral aspirant at the University of Kansas. He has a variety of interests in American politics and public policy and has coauthored a Southwestern Political Science Association paper (with Ken Collier) on the presidencies of Andrew Jackson and Ronald Reagan. He is a native of Omaha, Nebraska.

Robert A. Beauregard is Professor in the Graduate School of Public and International Affairs at the University of Pittsburgh and editor of *Economic Restructuring and Political Response* (Sage, 1989).

John J. Betancur is Assistant Professor of Urban Planning and Latin American Studies and a Research Assistant Professor at the Center for Urban Economic Development at the University of Illinois at Chicago (UIC). He has a Ph.D. in Public Policy Analysis from UIC and previously has taught urban planning in Medellin, Colombia, his native country. He conducts research on economic and community development strategies and on the experiences and conditions of Latino immigrants to the United States. He also has been active in Chicago community affairs for 15 years.

315

Richard D. Bingham is Professor of Public Administration and Urban Studies and Senior Research Scholar at the Urban Center, Cleveland State University. His latest books include *State and Local Government in a Changing Society* (with David Hedge; McGraw-Hill, 1991), *Financing Economic Development: An Industrial Response* (Sage, 1990, edited with Edward W. Hill and Sammis B. White), and *The Economic Restructuring of the American Midwest* (Kluwer, 1990, edited with Randall W. Eberts). He is coeditor of the journal *Economic Development Quarterly*.

John P. Blair received his Ph.D. in economics from West Virginia University and is currently Professor of Economics at Wright State. He has published widely in the field of economic development in such journals as *Urban Affairs Quarterly, Review of Regional Studies,* and *Economic Development Quarterly*. His latest book is *Urban and Regional Economics* (Irwin, 1991). He has served as a consultant to businesses and governments and is a member of the Montgomery County Planning Commission.

Barry Bozeman is Professor of Public Administration, Affiliate Professor of Engineering, and Director of the Technology and Information Policy Program at the Maxwell School at Syracuse University.

Daniel Bugler is Senior Research Associate in the Technology and Information Policy Program and doctoral candidate in the Department of Public Administration at the Maxwell School at Syracuse University.

Joan Fitzgerald has a joint appointment with the Centers for Urban Economic Development and Urban Educational Research and Development at the University of Illinois at Chicago. Her research focuses on how to better coordinate urban economic development with education and training. She is working on a research project, funded by the Joyce Foundation and Western Rural Development Center, that evaluates implementation of the Carl D. Perkins vocational legislation in Illinois. Recent articles appear in *Economic Development Quarterly, Environment and Planning,* and *Urban Affairs Quarterly*.

Douglas C. Gills is Assistant Professor of Urban Planning and Research Assistant Professor at the Center for Urban Economic Development at the University of Illinois at Chicago. He has a Ph.D. in political science from Northwestern University and has taught African American studies at the Peoples College in Chicago. He formerly was the Executive Director of Chicago's Kenwood-Oakland Community Organization. His research interests are community planning and economic development in low-income communities, political empowerment, and military conversion.

Robert Giloth is Executive Director of the Southeast Community Organization/Southeast Development, Inc., in Baltimore and has taught at the University of Maryland and at Tuft's University. Previous positions include Deputy Commissioner of the Chicago Department of Economic Development and Executive Director of the Eighteenth Street Development Corporation in Chicago.

William Woodbridge Goldsmith teaches at Cornell University. He is Director of the Program on International Studies in Planning and past Chair of the Department of City and Regional Planning. A noted specialist on urban policy in the United States and regional development in Latin America, he has lectured, researched, and consulted in the United States and abroad. His book *Separate Societies: Poverty and Inequality in U.S. Cities* (coauthored with Edward J. Blakely) was published in 1992.

Harvey A. Goldstein is Professor and Director of the Ph.D. program in the Department of City and Regional Planning at the University of North Carolina at Chapel Hill. His research interests and publications are on the role of scientific activity in generating regional economic development, methods of economic impact assessment, and the effect of regional economic restructuring in interpersonal income equality. His most recent book (with Michael Luger) is *Technology in the Garden: Research Parks and Regional Economic Development* (UNC Press, 1991).

C. Scott Holupka is Research Associate at the Vanderbilt Institute for Public Policy Studies, where he is part of the research team evaluating the Robert Wood Johnson/HUD Homeless Families Program. In addition to various applied research efforts, his work has also focused on community development issues, including an examination of the role that financial institutions play in neighborhood development.

Marie Howland is Director of the Urban Studies and Planning program at the University of Maryland, College Park. She is the author of *Plant Closings and Worker Displacement: The Regional Issues* (Upjohn Institute, 1988) and a forthcoming book with Amy Glasmeier titled *Rural Services in the Information Age*. She is also the author of articles on the relationship between national economic growth and the economies of cities and regions, urban national policy, rural economic development, and worker displacement.

Michael I. Luger is Associate Professor of City and Regional Planning and Chair of the Curriculum in Public Policy Analysis at the University of North Carolina at Chapel Hill. He has written extensively on regional economic development and technology policy and serves on the editorial board of *Economic Development Quarterly*. His current book project (with Harvey Goldstein) is on universities, technology development, and regional economic growth in the United States.

Julia Melkers is Assistant Professor of Public Administration in the School of Business and Public Administration at the University of Alaska Southeast.

Robert Mier is Professor of Urban Planning and Public Administration at the University of Illinois at Chicago. He is widely recognized for his policy research and consulting in the areas of economic development strategies and for his teaching and professional training in public systems management. He is the author of *Social Justice and Local Development Policy*. From 1983 to 1989 he was Director of Development for the City of Chicago during the tenure of Mayors Harold Washington and Eugene Sawyer. He was the architect and chief implementor of Chicago's highly regarded 1984 Development Plan, a national model for equity-oriented local municipal development planning. He has degrees in civil engineering from the University of Notre Dame and urban planning from Cornell University and is a registered professional engineer. He also is a decorated veteran of the Vietnam War and an avid, though aging, rugby player.

Arthur C. Nelson is Professor of City Planning, Public Policy, and International Affairs at the Georgia Institute of Technology. He is consultant to the Economic Development Administration and economic development agencies in several states and major cities. He has pioneered research in rural enterprise zones and speculative industrial buildings. His most recent economic development work is in exurban industrialization. He serves as an editor of the *Journal of the American Planning Association* and associate editor of the *Journal of Urban Affairs*. He is widely published in the areas of regional development planning, urban form, resource land preservation, and infrastructure planning and financing. He was director of planning and coordinator of economic development for an exurban county in Oregon and was a consulting development planner along the West Coast during 1972 to 1984. He earned his doctorate in urban studies from Portland State University concentrating in regional science and regional planning. Prior to his appointment to Georgia Tech, he served on the faculties of Kansas State University, the University of New Orleans, and Southern University.

Robert Premus is Professor of Economics at Wright State University. A former economist with the U.S. Joint Economic Committee, he has completed numerous studies on regional development and industrial location. A frequent consultant to government and business leaders, his economic policy studies on entrepreneurship, venture capital, and regional development have received national recognition. He is the principal author of numerous economic studies for the U.S. Air Force.

Lewis A. Randolph is Assistant Professor in the Political Science Department/Public Administration Program at Ohio University. He received his Ph.D. in political science at The Ohio State University. His areas of interest are urban development, urban politics, public policy, black politics, and American politics. He is a coauthor with Rob Mier and Joan Fitzgerald of "African-American Officials and the Future of Progressive Political Movements" in *Local Economic Development Policy Formation: Experiences in the United States and United Kingdom* (edited by David Fasenfest; St. Martin's, 1992).

Elaine B. Sharp is Professor and Chair of Political Science at the University of Kansas and President-Elect of the Urban Politics Section of the American Political Science Association. She is the author of *Urban Politics and Administration* (Longman, 1990), *Citizen Demand-Making in the Urban Context* (University of Alabama Press, 1986), and a number of articles on economic development, citizen participation, and urban affairs. She is a native of the Pittsburgh, Pennsylvania, area.

Anne B. Shlay is Associate Director of the Institute for Public Policy Studies and Associate Professor of Geography and Urban Studies at Temple University. Her research has looked at the political and social forces governing urban development including zoning, housing and neighborhood ideology, growth coalitions, and financial institutions. In addition, she studies problems of acute poverty and homelessness.

Michael Teitz is Professor of City and Regional Planning at the University of California at Berkeley. He teaches, does research, and consults on housing and economic development in the United States and abroad.

Wim Wiewel is Director of the Center for Urban Economic Development and Associate Professor in the School of Urban Planning and Policy at the University of Illinois at Chicago. Through the center, he provides technical assistance and research in the area of economic development to community organizations and local governments.